PRACTICAL GUIDE TO SOLAR HOMES

PRACTICAL GUIDE TO
SOLAR HOMES

by the Editors of
Hudson Home Guides

BANTAM/HUDSON IDEA BOOKS
New York, New York • Los Altos, California

PRACTICAL GUIDE
TO SOLAR HOMES.
A Bantam Book/published in
association with Hudson Home
Publications, February 1978

Executive Editor, Robert J. Dunn;
Book Editor, Sandra L. Beggs;
Project Editors, Phyllis Askey, David
Edwards, Christina Nelson, Joseph F.
Schram; Solar Consultant, Michael J.
O'Hearn; Art Director/Designer,
Annette T. Yatovitz; Illustration/
Graphics, Carol Johnston, Barbara
M. Thompson; Architectural Render-
ings, Kenneth Vendley; Production,
Richard Osborne, Carolyn M.
Thompson; Composition, Ulla
Wahlin, Phyllis Woodbury

Cover Rendering, Kenneth Vendley;
Solar Home Design, Total Environ-
mental Action, Inc. of Harrisville,
New Hampshire.

ISBN 0-553-M01132-4
Published simultaneously in the
United States and Canada

Bantam Books are published by
Bantam Books, Inc. Its trademark,
consisting of the words "Bantam
Books" and the portrayal of a
bantam, is registered in the United
States Patent Office and in other
countries. Marca Registrada. Bantam
Books, Inc., 666 Fifth Avenue, New
York, New York 10019.

Printed in the United States of
America. Library of Congress catalog
card number 77-91064.

Contents

Preface

The idea of harnessing the sun's energy is not new to mankind. Man has been doing it intuitively for eons of time in order to improve some facet of his living environment. However, it is only within the past 50 years that solar energy as a source of home comfort has been with us as a technology. Some very basic inroads toward improving the technology were made in the 30's, 40's and 50's, but the efforts dropped off sharply in the belief that we were a nation — a world — of plenty in terms of our energy resources. A variety of crises in the late 60's and early 70's brought the subject into acute focus once more and the realities of finite energy resources once again to the fore. As a direct consequence the idea of capturing the infinite capacities of the sun as a fuel for home comfort has become a RENEWED technology.

There are countless books in print purporting to explain this renewed technology in varying degrees of depth. Most, on review, turn out to be much too technical for the average homeowner. It is our objective with this book to make the basic principles of solar energy and other forms of energy conservation both readable and understandable to the layman. We do not look on solar energy as it applies to the home environment as a do-it-yourself project for the typical homeowner. We assume that most homeowners will be seeking professional assistance. Consequently, it is our objective to make this book a primer so that the average homeowner will become sufficiently conversant with the subject to communicate with a professional in the field.

ACKNOWLEDGEMENT
The editors wish to acknowledge the information in Chapter 2 adapted from **Solar Dwelling Design Concepts**, published by the U.S. Department of Housing and Urban Development (HUD), Office of Policy Development and Research, May 1976; material created by AIA Research Corporation, Washington, D.C., contract 1AA H-5574.

Photography: Sterling Roberts and Steve Ford

Solar home in northeastern Ohio uses water-to-air heat pump with solar heated water for a forced air system. See page 68 for more about this solar home.

CHAPTER 1

Solar... Energy of the Future

Energy crisis? Doom? Real? Imagined? Does it really make any difference? Ask those questions to the experts and there will be as many different answers as persons asked, slanted of course, toward their areas of concern. It remains, however, that the American people were made acutely aware of what an energy crisis could be like when oil from the Mid East was embargoed in late 1973-74. And there have been other symptoms sampled when entire American cities suffered blackouts. The ensuing picture, the public response, was anything but pleasant.

Can these short term dilemmas be reckoned as a sign of things to come? We don't think so. The need to adapt or change can be a positive thing, and we prefer to think that the American people are an adaptable lot. But these mini-crises, when added to a few dedicated voices that have been decrying our utter waste of energy in general, and the finite fund of fossil fuel in particular, can be a positive thing.

In this case it was positive because it sent a techno-logical nation back to its collective drawing boards in a desperate quest to reclaim a solar technology that had been conceived literally hundreds of years ago, nurtured to an experimentally healthy fetus stage and then aborted for the most part by a nation steeped in seeming abundance.

So now that we've seen the blinking red light of total drain casting its ominous warning glow on the not-too-distant future of existing resources, we, as a nation, have begun casting about for yet other sources. They are there, there is no question about it. Who can foresee an end to the interaction of sun, wind, atmosphere (no matter how acrid we seem to be making it), and the oceans? All are seemingly indomitable forces in a sense of total control. But the quest is not control. The quest is to discover a variety of ways to harness and employ these basic interactions as diverse sources of energy at a cost that will approxi-mate what each in its own sphere of the elements costs us to experience now.

It has already been established that at some time in the future our fossil fuel resources will come to an end. When, nobody really seems to know and it's for certain that all won't run out at the same time. However, as one is depleted, emphasis tends to be put on another,

and it could be that the last gasps could begin very early in the 21st century.

It would seem that we, as a nation, have a dual option. Our efforts at conservation must be expanded in order to keep the end of what we have as far at bay as possible. There are myriad ways to accomplish this, but it is imperative that we have a governmental leadership capable of educating and monitoring the populace to insure we're aware of all conservation possibilities and that we play by the rules of the game . . . with adequate penalties if we don't.

It is a distinct possibility that such strong leadership will evolve, but to date it has not. There has been lip service paid to the need, and agencies established that seem to be sparring with other already estab-lished agencies, rather than moving ahead. Priorities must be established. And it would seem that the prospect of a nation devoid of energy resources would be sufficient impetus. Unfortunately we don't antici-pate any weakening in the vested interest population, so we can only hope for potent leadership.

On the other hand, funds must be made available for the more rapid development of our infinite resources. Monies have been nominal to date. It is conceivable that if this is accomplished, in sufficient time, the drain on the fossil fuel supply will lessen, making our down-the-road alternatives broader yet.

The marketplace is rampant with books about alternative sources of energy. Many are technically oriented and some, ostensibly directed at the average homeowner whose primary concern is the energy support systems for his own family, are still a bit heavy on the technical side. There remain, however, some really pertinent books . . . all with a sincere purpose.

This book is expressly for the average homeowner. Its purpose is dual. It will stress the unqualified need for energy conservation in home building, and illustrate the latest inroads made in the field of solar energy in the past few years . . . miniscule as they may be in terms of a panacea. Heating and cooling a home with solar energy — as well as solar for cooking, heating water and pools — is already upon us as a workable alternative. However, the state of the art is definitely in its infancy. We will show you that state of the art and encourage you to learn the rudiments well.

The Sun...
The Source

The sun is the star around which the earth revolves and it is our planet's source of light and heat. When you come right down to it, it is the light and heat from the sun that make this earth of ours habitable at all.

It is, in fact, the ultimate source of nearly all energy utilized by industrial civilizations in the form of water power, fuels and wind. Only atomic energy, radio-activity and the lunar tides are examples of non-solar energy.

All forms of life depend on the sun and its electro-magnetic rays. Millions of years ago, chemical reactions stimulated by the sun formed all fossil fuels, and without this process, there would have been no such thing as coal, natural gas or petroleum. However, these fossil fuels, as we have already noted, are finite, costly and getting costlier.

Of these fossil fuels, coal is relatively the more plentiful. Just after the mid-19th century, up till when wood was a primary source of fuel, coal accounted for only about 10 percent of fuel for the home. Its usage during the next half century increased rapidly to the point where, in 1910, it was approximately 75 percent of the home fuel market. Gradually, as other forms of fuel, oil and natural gas were found to be employ-able coal with its sooty, polluting fallout lost its toe-hold on the market. Today, as a home fuel, it is right back to where it was in 1860, just about 10 percent. However, with the threat of rapidly depleting oil and natural gas resources, we shall probably see a comeback of coal as a home fuel with, of course, an improved technology targeted at minimizing its pollution factor. Our ultimate aim, however, must be to "capture" the infinite energy sources offered to us directly by the sun... and infinite it is.

This potent star is firmly ensconced some 93 million miles from the earth, its position changing by the season, day, and hour in relation to our planet, yet we can chart its whereabouts accurately. Earth receives about one-half of one-billionth of the total energy output of the sun. This tiny fraction of this gaseous orb's efforts represents some 745 million billion kilowatt hours annually on earth . . . and the natural life of our planet uses only a minute portion of that energy. Mankind on this earth uses considerably more, and that usage is increasing exponentially.

The sun retains 35 percent of its own energy in a self-regenerating process. Of that energy directed toward the earth, 50 percent is reflected by the atmosphere before reaching the surface, 15 percent is reflected by that surface, 5.3 percent is absorbed by bare soil and the remaining 29.7 percent is absorbed by the hydrosphere to raise water temperature and eva-porate water, and is absorbed by land and marine vegetation.

Considered theoretically, the sun's rays hit the earth's atmosphere at just about 444 BTU's (British thermal units . . . equal to 130 watts in electrical power units) per hour for every square foot of area on earth. Considering atmospheric absorption, hours of dark-ness, cloud cover and geographical location, the average amount of sunlight falling on a square foot of ground in the United States, on a year-round basis, is about 13 percent or 58.5 BTU's per hour (the equivalent of 16.4 watts in electrical power units). This amount of heat, if absorbed and retained by a little less than a cubic foot of water, would raise the temperature of the water by 1° Fahrenheit in the course of an hour.

However, that is theory. There are, as noted, any number of variables that must be taken into account in this imperfect system of the universe — imperfect only in the sense of our hypothesis — and there are such things as cloud cover that diminish the energy potential, nighttime for obvious reasons, and the seasons. From such variables does the real challenge arise. How do we collect and store this infinite fund of energy for controlled dispersion when and where mankind's needs may arise? And, of course, do it

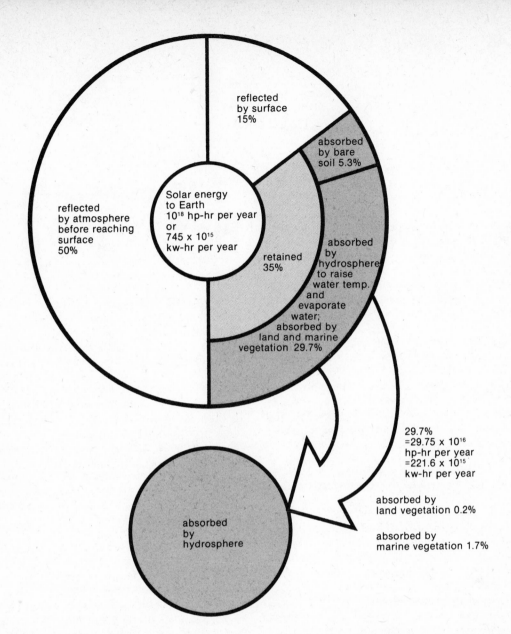

Solar energy
to Earth
10^{18} hp-hr per year
or
745×10^{15}
kw-hr per year

reflected
by surface
15%

absorbed
by bare
soil 5.3%

reflected
by atmosphere
before reaching
surface
50%

absorbed
by
hydrosphere
to raise
water temp.
and
evaporate
water;
absorbed by
land and marine
vegetation 29.7%

retained
35%

29.7%
=29.75×10^{16}
hp-hr per year
=221.6×10^{15}
kw-hr per year

absorbed by
land vegetation 0.2%

absorbed by
marine vegetation 1.7%

absorbed
by
hydrosphere

efficiently and economically.

Where, one asks, is the hang-up? There have already been some excellent inroads into solar collecting technology. Few are either totally efficient or economical . . . and that is the real crux of the problem. Many systems across the nation, for example, have been fabricated from materials already existing on the market today. But these are not necessarily low-cost products and relatively few individuals have the confidence or technological where-withal to do it themselves. Plastics, for example, are a natural for solar applications of the future, both transparent and reflecting plastics. Yet, the material is not particularly inexpensive, readily available for easy adaptation to solar technology and little is generally known (by the population at large) of its durability in a constant exposure situation. At this point in time, the general dissemination of information on this subject is just about nil, even as it relates to the state of a less than perfect art.

Solar cells, photovoltaic cells, are also available but hardly for a mass market in terms of cost. While we agree that it is difficult to report on an evolving subject, we do feel that periodic reports, emanating from reliable government sources on materials and devices such as these, are terribly important in terms of a national state of confidence that something is being done with respect to the energy concerns of an ever-expanding populace.

One other kind of information that is important for the homeowner to have at his fingertips is that which explains the solar energy factors as they relate to specific geographical areas of the country. What is the average sun situation? What kind of cloud cover can be expected? Average temperatures of the season? What average periods of time is the area without sun at all? In other words, is solar energy practical in that specific area as it is in its present state or, in other words, would a solar converting system be efficient and economical? This kind of information could easily and should readily be available and freely disseminated for every section of the country.

Is Solar Practical for Heating, for Cooling, for Hot Water?

Each individual who has an inkling that solar energy might be a logical source of energy supply for his home must review the prospects from his own standpoint. The reality of the situation is that solar energy as a source of energy for heating and cooling the home, generating hot water and even, in some instances, cooking, can be the way to go. However, the variables affecting individual families across the United States are infinite and require an absolute objective evaluation to establish its practicality. The following are some of the more pertinent questions that should be included in that analysis.

Q Is my present home as energy efficient as possible?
A The fact is that most homes in the United States are considerably less than energy efficient and, as a result, require a great deal more fuels to keep them comfortable than is really necessary. The efficiency aspect of any home includes all aspects of a home, from the planning stages through the selection of energy-saving appliances and other products. (For an excellent example of a house constructed with total energy conservation in mind, see TERA One, pages 16-19.) Not every existing home can be made totally energy efficient, but many defensive measures can be added that will substantially reduce present operating costs, perhaps even eliminating the need for a total solar system once the cost factors are analyzed after the improvement is made.

Q What can I do to assess my current home situation?
A We strongly recommend professional assistance because of the many considerations that must be sorted out. There are numerous books on the market that can help you calculate your needs in terms of energy conservation and/or solar potential, but it is, at best, difficult to go it alone. Most architects or solar engineers would be capable, but we strongly recommend that you get more than one assessment, particularly if the assessor has a sales interest in your situation. If you are considering a new home, we feel you should seek out an architect who has already established a working relationship with a solar firm or a solar engineer. It is imperative, too, that the home builder be familiar with solar systems. Without the working relationship, costs could become astronomical, making payback impractical.

Q Is it practical to think in terms of a solar system as a total heating source for my home?
A In effect, it is possible, but we do not think the solar "art" has evolved sufficiently for you to think in those terms. Throughout the industry, you will turn up claims for systems that can supply some 80-90 percent of your energy needs. For solar-generated energy for heating, cooling and hot water, we feel that you would be safer to think in terms of 50-60 percent savings on your fuel costs. You may, in the long run, do better, but remember, most calculations involving how

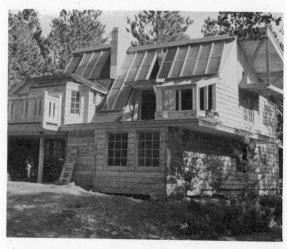

A solar retrofit. Solar Power Supply.

much sun your system will receive in order to generate energy are based on average figures for your particular geographical climate, so the best calculations you can get are really probable. We feel that it is best that anyone thinking solar at this point in time really consider an adequate backup system that could handle the total energy needs of the home. It can be very practical to think of solar only in terms of your hot water supply, which represents a goodly portion of your utility bills. Typically, the installation cost (or hookup to an existing system) is much less than a system designed to generate heat for an entire home and hot water probably represents the biggest waste factor in the American home. Generally speaking, it is something we have always taken for granted, and seldom is the cost considered.

Q Is cooling a home with solar energy possible?
A The heat from a solar collector can be used to drive an air conditioning system. Although a variety of systems are possible, the most popular to date is the absorption-refrigeration method. It is a very complicated process and, as such, is very expensive to install. The physical principle that applies is that evaporation is a cooling process and condensation is a warming process. When liquid changes to gas, it is called evaporation and when gas becomes liquid it is called condensation. This cooling-warming cycle happens several times in a solar air conditioning system. The cool air resulting from evaporation is sent into the house and the warmer air is passed to the outside. In every such process, energy is lost and the heat from the sun makes up for this lost energy, just as electricity or natural gas make up for energy lost by standard refrigerators and air conditioners.

Q Assuming solar is economically practical, is it worth it in the long run?
A The answer to that is yes, with reservations. If you are able to have an excellent system installed — one that is guaranteed to be trouble-free, need minimum maintenance and last 20-30 years — at a reasonable cost, you will have made the right move . . . even if your real savings on your fuel bill are only 50-60 percent. Some 31 states, as of this writing, have tax deduction legislation in effect for solar installations, and it is an almost certain thing that the Federal government will initiate tax incentives relative to energy-saving solar installations. What is more, we feel that solar systems will rapidly become a major selling factor in residential home building in the near future. So, if your system is a good one, and you plan to sell your home down the road, you may be able to demand a premium price for your home.

Q Is is possible to add a solar system to an existing home?
A Yes, but there are many circumstances where it is not at all practical or economical. An existing home may require a roof alteration or other construction in order to gain the proper, most efficient tilt for the solar collectors. This can be costly, and if the home does not have just the right orientation to the sun for ideal solar collection, then additional collector units must be added, again driving up the cost of the system. In terms of add-on efficiency where a solar system is concerned, a solar hot water system is practical just about anywhere in the country. Everyone uses (and wastes) hot water. On the contrary, adding a solar home heating system in parts of the country where fuel demands on an annual basis are minimal is not particularly financially sound, although it makes good sense in terms of conserving a dwindling energy supply. The cost factors and home design prerequisites for the installation of a solar hot water heating system are considerably less than tackling the overall heating needs of a home. Less collector capacity is needed and, as a result, collector positioning is less of a problem. So, adding solar to an existing home is very much a possibility, but study the economics very well. You may find that increasing the energy efficiency of the home through other methods will suit your needs (and pocketbook) better.

Q Is solar still too new to try?
A No, we don't think so. The use of the sun as a source of heat has been with us for ages. Initially, it was intuitive as in the case of the Pueblo Indians of the southwest, who lived in a hot arid climate. Their days were hot, and their nights were cold. Without technology, they built their homes of adobe, a building material with high heat absorbency. Their adobe walls would absorb the hot rays of the sun during the day and re-radiate them into the dwelling's interior as the night chill became a factor. In the 1930's, Chicago architects George and William Keck began designing homes with entire glazed walls facing south and with heat-absorbing materials inside that would function as a re-radiating medium. There were numerous experiments with solar energy for home environmental control through the 1940's and 1950's, but these were pretty much individual efforts or academic research (Massachusetts Institute of Technology's four successive efforts are the most notable). Funds for continuing research were not readily available and relatively little was done to expand the technology until the early 1970's. And even then it took a succession of crises to bring the solar energy potential back to the surface. Given the world energy situation, and the technological inroads made in the past few years, it is a certainty that solar energy for the home will grow exponentially over the next ten years, to a point where good, solid, all-encompassing and long-lasting systems are readily available for the low-cost comfort of the American populace at home. If you do "go solar," go slowly and be sure to investigate any companies or manufacturers with whom you will be working, thoroughly.

Energy Efficiency with Solar Potential

Early man intuitively knew how to use solar energy to heat and cool his living environment. He realized the advantages of placing wall openings in the direction of the sun's path to capture its warmth during winter days. He discovered the unique ability of certain materials to retain that warmth and release it later after the sun had set. Man's search for natural efficiency continued and, indeed, increased as he became part of a tool and craft society. Reacting to the present energy crisis, man is again emphasizing this age-old concept, applying energy-saving ideas to building

design. There are a number of very fundamental techniques which can be incorporated into home construction to achieve an energy-efficient structure, thus decreasing our reliance on a dwindling supply of fossil fuels. One way to accomplish this is to use the local climate and topography to reduce dependence on mechanical energy; another is to thoroughly insulate the home throughout and provide excellent ventilation. The typical home design shown below illustrates some basic techniques a homeowner can employ to obtain an energy-efficient dwelling.

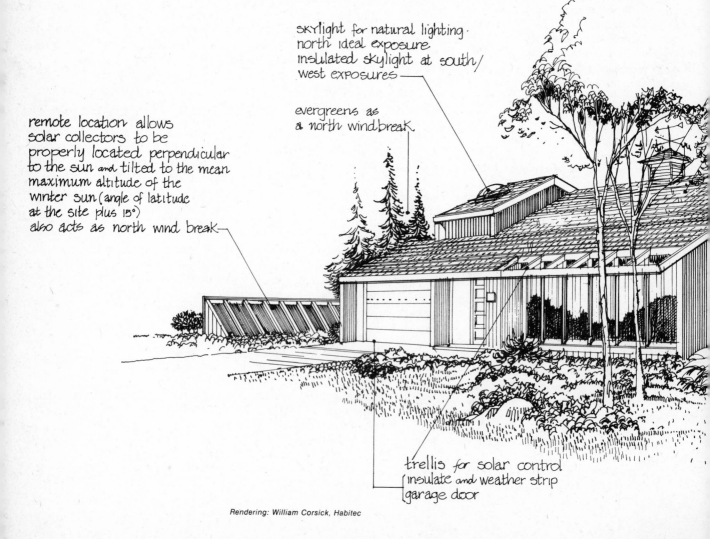

skylight for natural lighting. north ideal exposure. insulated skylight at south/west exposures

evergreens as a north windbreak.

remote location allows solar collectors to be properly located perpendicular to the sun and tilted to the mean maximum altitude of the winter sun (angle of latitude at the site plus 15°) also acts as north wind break

trellis for solar control insulate and weather strip garage door

Rendering: William Corsick, Habitec

a belvedere lets hot air escape in summer
letting cool air come in the house's windows more readily.
light color roofing · reflects heat
attic vent - expels unwanted hot air and potentially harmful moisture.
(insulate attic)
leaf bearing trees on south/west exposure offer shade in summer
and in winter let sunlight through their bare branches for warmth.

insulate exterior walls

reflective glass or coated glass to minimize
summer heat (south/west exposure.)

low level vents let in cool air from shaded ground areas
or cool night air enter the house in summer.
(insulate crawl space.)
caulk and seal around chimney

vegetation as a sound barrier and
air purified

Photographer: Steve Marley

Energy Efficiency with a Solar System

The home displayed on these pages is called TERA One and it is, in effect, an experiment in Total Energy Resource Application. Conceived by the Pacific Power and Light Company of Portland, Oregon, the design and concept were developed with the architectural-engineering planning firm of Skidmore, Owings & Merrill. The project was built on a site belonging to the City of Portland. Much of the material, land and instrumentation was donated by manufacturers and interested organizations (see appendix). No government grants were used.

While it is an experiment, it could also comfortably house a family of four. It is a "solar house," but it goes considerably beyond that. It is in many ways the house of the future. One where the design has overcome many of the design flaws so rampant in home building throughout the nation. In effect, TERA One has been designed to work overtime to reduce the overwhelming amount of energy wasted in the typical home, and yet preserve the visual aesthetics and total comfort we expect in a new home today.

The dominant feature of TERA One's exterior appearance is the steeply sloping solar collector wall on the structure's south side. Emphasis on energy conservation is reflected throughout the home's design and landscaping.

16

• The interior arrangement of space is designed to minimize space heating requirements
• Building surface is reduced through partial two-story construction
• There is an extensive use of natural wood windows to maximize natural light input
• Greenhouses, pleasing to the eye, become an integral part of the heating/cooling system
• The vestibule at the front door on the north side reduces outside air infiltration
• Deciduous (drop leaves) trees are planted on the south side to shade the house during hot months and allow the sun in during the winter months

The home is designed to utilize solar energy for heating whenever possible. The solar collector is on the south side of the house to gain the greatest solar radiation. It works in a simple fashion.

Basically, the collector is a thin rectangular box. The top is composed of two sheets of glass, with a one-half-inch air space in between. The sides and back are insulated and the inside is made of metal which has a special flat black coating to increase solar absorption. This is called the solar plate.

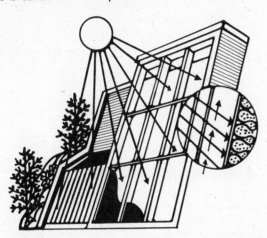

On this house, 185 square feet of active collector plate is used to "power" the heating system. When sunlight strikes the blackened surface, heat is absorbed from the solar radiation, which in turn warms the air in the enclosed collector box. The heat is then drawn off by mechanical fans which move the air inside. Heat from the collector on TERA One can be used right away, or, by means of using a heat exchanger in the heated air stream, to warm tanks of water in the basement. TERA One has a backup electrical heating system for periods when there is inadequate solar radiation.

INSULATION

Several methods of insulating a home have been incorporated into this project and in several important aspects they have been improved upon.
• Oversized wood members are used for most exterior wall framing. The space this provides has then been filled with loose bulk insulating material.

• Other walls are insulated with mineral blanket or foam, so that all the methods can be compared.
• Earth is employed as an insulating material by building it up against exterior walls and on top of a portion of the roof.
• The foundation is insulated with foam board.

KITCHEN

All appliances — refrigerator, washer-dryer, dishwasher and range — have been selected on the basis of energy efficiency. A heat reclaiming system is used to extract waste heat from the dishwasher and clothes washer. This potentially wasted heat is then used to preheat the domestic hot water.

HEATING & COOLING

The heating and cooling system serving TERA One is a combination of mechanical and natural systems that attempt to get the most energy possible from the sun.

In this particular project, the storage medium for unused heat is water, which is contained in large insulated metal tanks in the basement. The device which transfers the energy from warm air to water is the primary heat exchanger. It is basically a finned coil through which water flows. Air passes over the fins and the heat is transferred from the air to water. When the direct solar input is less than necessary to heat the home, the system draws the heat from the storage tanks . . . again this is done through the primary heat exchanger, and the warmed air is then pumped into the home.

When the stored heat is used up, then electrical energy is required. In order to extract the maximum energy from the storage tanks, a very efficient electrically powered device called a heat pump is used. This heat pump has the ability to extract residual heat from cold water to produce hot water. But the heat pump cannot operate indefinitely pulling heat out of the cold tanks. When it reaches its mechanical limit, a six kilowatt electric boiler is put into operation to continue heating the warm storage tanks.

SOLAR . . . ENERGY OF THE FUTURE

The home can be cooled in a number of ways. As for heating, the first capacity to be used is the natural one. The home is designed to make maximum use of natural ventilation when outside temperatures are acceptable. If natural means to cool the home are inadequate, then the system again turns to the heat pump. The heat pump can also produce cold water, which circulates through the primary heat exchanger resulting in cool air which is blown into the home. The air ducts that carry the conditioned air throughout the home are heavily insulated on the inside to keep heat or cold in and minimize noise.

During the summer the greenhouse will be part of the cooling system. At night, plant transpiration and water evaporation will provide a source of cool air that can be used in the house during the next day.

FIREPLACE

The fireplace for the project looks like any other one, except it is energy efficient. It will not use inside air combustion. The air inside the house is after all, already warmed. To use it for combustion would be wasteful as it would simply be drawn up the chimney and dumped outside. The combustion air used in the fireplace is drawn into the firebox through a ducted opening to the outside. The glass screen in front of the fireplace prevents room air from being utilized in the combustion process.

Heat from the fireplace is conducted into the living room by radiation through the glass screen and from a heat exchanger located in the flue. Cool room air is drawn into two ducts located at each side at the base. Small fans push the room air through the heat exchanger where it is warmed by the fire and conducted back into the living room.

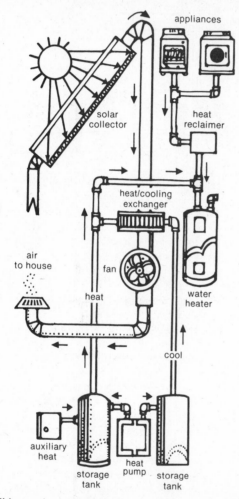

NATURAL LIGHT

Windows are, in effect, very inefficient heat insulators. TERA One windows are planned to minimize heat loss while maximizing the amount of sunlight. All windows are double glazed to decrease the flow of heat through the glass. In some locations, openings occur within the house to allow light from one area to spill into another area.

NATURAL VENTILATION

Natural air flow is a specific design feature that is often overlooked in modern housing, but it was well recognized in the early days. TERA One is designed with

Woven throughout this simplified description of TERA One's heating and cooling design is a sophisticated control system that integrates all of the various systems.

GREENHOUSE

The south greenhouse is both an aesthetic and functional feature. During the winter months, the enclosed space becomes an additional heat gathering source . . . heat which will be drawn inside the home to supplement the system by means of gravity ventilation openings.

ventilation openings on north and south sides to permit the entry of outside air when necessary for cooling.

As mentioned earlier, TERA One is a test house. An extremely sophisticated computer and data logging system keeps track of: solar radiation available; wind speed and direction; outside temperatures for air and ground; inside temperatures for walls and space; mechanical system temperatures for collector, storage, air in and out; heat flow through the walls to evaluate insulation systems; performance of mechanical systems in terms of use, capacities, collector gain, status.

However, for the purpose of this book, it reflects what can be done to make a home energy efficient in conjunction with a solar system. We have included a cost break-down of the basic construction process and then further broken down the additives for the experimental process. Over and above the experimental process, we feel there are numerous adaptable ideas which can benefit the energy conscious homeowner.

ESTIMATED CONSTRUCTION COSTS

(These approximate costs are typical of construction and equipment during the construction period of TERA One. These costs should not be construed as firm costs in the face of an inflationary market but are provided for general information only).

1. 1250 square feet of floor space at $25/sq. ft. $43,800

ADDED COST FOR THE AMENITIES UNDER TEST AT TERA ONE

2. Vestibule.......................... $3,000
3. Greenhouse 1,500
4. Plant Material 1,000
5. Solar Collector 2,700
6. Reflective Deck................... 5,000
7. Mechanical and Storage Systems 3,800
8. Control System 5,000
9. Insulating Glass 2,500
10. Extra Insulation 900
11. Sealing & Weatherstripping 500
12. Louvers.............................. 1,000
13. Concrete and Wall Variances from Standard 5,000
14. Shade Trees 200

$32,100

TOTAL $75,900
($60.72/sq. ft.)

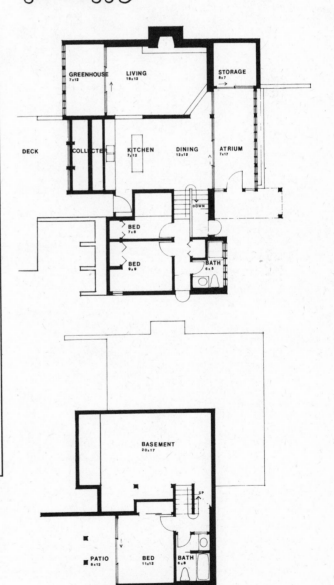

Design: Earth Dynamics, I...

Design: Solar Power Supply

Design: Energy Systems, Inc.

Design: Burt, Hill & Associates, Architects

Design: Copper Development Association, In...

Design: Portland General Electric Co.   Revere modular solar units

CHAPTER 2

Planning Your Solar System

The solar industry has been going through a reincarnation process and that bodes well for the homeowning public. Solar energy as an alternative source of energy in the home is a genuine concept. It remains, however, an evolving science. It could be your solution to outrageous fuel bills and even more outrageous consumption of limited fossil fuels. It is important, if you are considering the use of some sort of solar energy, that you be totally aware of the state of the art and how that state of the art relates to your specific living environment before you decide.

We can provide you with the basics here in this book, and then go on to show you how others have applied the solar science to their environment. If, after conducting your own investigation you feel solar is right for you, we strongly suggest that you review your thinking with a professional in the field . . . an architect or solar engineer in your area who understands all of the technical vagaries of the subject.

The object in considering solar at all, of course, is to gain the optimum in both comfort and savings in the home. Probably the most important factor you must be aware of is the climatic/geographical situation as it relates to your particular site. These are given factors over which you have no control, but they, in turn, exercise a tremendous control over whatever potential you have for the utilization of solar energy.

This includes such things as the geography of the area surrounding the site, its topography and orientation of the slopes, the underlying geology, soil potential (and constraints), vegetation, areas exposed-protected, natural access areas to and through the site, and prevailing patterns of solar radiation, wind, precipitation, temperature, as well as water and/or air drainage patterns.

As far as solar systems are concerned, there is a great deal of discussion throughout the solar industry regarding definitions of active and passive systems. In reality there is no black and white definition. In effect, a passive solar heating and cooling system is one in which the thermal energy flows by natural means, while an active system utilizes man-made hardware to enhance the collecting, transfer and distribution facets of the sun's energy into home comfort.

Throughout this section of the book, we will explore some of the more pertinent systems used in residential situations in their current state of the art. There really is no better or best system, since all are relative to the specific location where the solar system is to be installed. There are, however, some basic differences that you should be aware of.

A passive system, for example, works 100 percent of the time whenever there is sunshine. It collects every bit of energy, direct or diffuse, that comes into the home. An active system, on the other hand, has a threshold that kicks it on and off, and consequently, it does not begin to work until a certain temperature is reached. Naturally, the passive system loses more energy to waste than an active system, which is a natural balance for its ability to work all the time.

One other consideration that should enter into your solar decision was covered in the previous chapter where we discussed and analyzed the economies possible in a new or remodeled home without any consideration of a solar system at all. Most any solar system you might be considering will involve an initial dollar outlay. Over a reasonable period of years, that outlay should have a payback factor that makes the installation costs worth your while. For example, if your situation is such that you are going to spend something in the vicinity of an additional $10,000 on the cost of a new home to add a solar system, you should review that expense in terms of potential savings calculations on fuel bills over the next few years, as well as an added estimated value to the house should you decide to sell in a decade or so. Fossil fuel predictions are dire right now in the seventies and there is no reason to assume they will improve. Will the situation be even more dramatic at the end of the eighties, in effect making a reasonably effective solar system extremely valuable? That certainly could be. And it is this type of projection that should help you analyze exactly how far you want to go with a solar system.

We suggest that anyone planning to make solar a viable part of his home comfort, not try to cut corners in any present installation. Solar technology does work and, at the rate developments are coming about the technology should improve drastically in the next 10 years. Be sure you take full advantage of all that's available to you now.

Climate...
The Given Conditions

Global climatic factors such as solar radiation at the earth's surface, tilt of the earth's axis, air movement as well as the topographical situation determine the climatic makeup of any area on earth. These are the determinants of temperature, humidity, solar radiation, air movement, wind and sky conditions for any specific location.

Naturally, regional factors play a role in the climatic situation where we find such things as mountains, bodies of water and open land lending their peculiar characteristics to a basic climatic condition. And then, as you move to a particular selected site, you will find that site influenced by such variables as site topography, ground surface, and three dimensional objects. All of these factors, not in and of themselves, but in combination are important determinants for selecting your own personal approach.

There are a variety of definitions for climate, but it stands to reason that climate is really the sum total of the weather as it appears at any particular locale. The climates of particular localities are comparatively constant, and despite pronounced and relatively unpredictable changes, they have predictable patterns that repeat themselves often enough to establish a common denominator for that area. And it is these common denominators that become the determinants for the climatic phase of solar selection.

Any designer or builder is primarily interested in the elements of climate that affect human comfort and, in the long run, the design and use of buildings. The basic information needed to make an intelligent building decision — and here we are referring to solar building — include averages, changes and extremes of temperature between day and night, humidity averages, how much radiation strikes the area (and how much is lost), which way the air is moving, when and with how much force. Snowfall and its distribution, sky condition varieties and other special conditions probable in the area, such as hurricanes, hail or thunderstorm activities, also play a determining role.

Generally speaking, these factors are regularly recorded by the National Weather Service, however, they are not recorded with home building in mind and will, of necessity incorporate much data unnecessary for home building needs. There are some helpful climatic design publications available which can be extremely beneficial and are relatively easily translatable to design needs. The following publications are the most informative regarding temperature, degree days, solar radiation and other climatic data:

- ASHRAE Handbook of Fundamentals, American Society of Heating, Refrigeration, and Air-conditioning Engineers, Inc., 1972.
- National Association of Home Builders (NAHB), Insulation Manual: Homes, Apartments, 627 Southlawn Lane, P. O. Box 1627, Rockville, Maryland 20850.
- U. S. Climatic Atlas, Government Printing Office, Washington, D. C., June 1968.

There have been a number of systems proposed for classifying climatic regions within the United States. Koppen's classification of climates, based on vegetation, has been the basics for numerous studies of the relationship of housing design and climate. Under this system, four broad climatic zones are found in the United States: cool, temperate, hot-arid, and hot-humid (see map). The climatic characteristics of each region are not uniform. They may vary both between and within the regions, and they are not as abrupt as indicated on the illustration. Each different climatic region merges gradually and almost invisibly into the next one. In fact, it is not unusual for one region to exhibit at one time or another the characteristics associated with all the other regions. However, each region has what might be called an inherent character of weather patterns that distinguishes it from the others. The following descriptions of the various regions offer some insight into the general conditions to which solar building design in those particular regions must be responsive.

COOL REGIONS: A wide range of temperature is characteristic of cool regions. Temperatures of minus 30°F to plus 100°F have been recorded. Hot summers with cold winters are typical. Winds, generally out of the NW and SE, persist throughout the year. And it is important to note that the northern regions which are associated with cool climates receive less solar radiation than southern locations.

TEMPERATE REGIONS: An equal distribution of overheated and underheated periods is characteristic of the temperate regions. Seasonal winds from the NW and S along with periods of high humidity and large

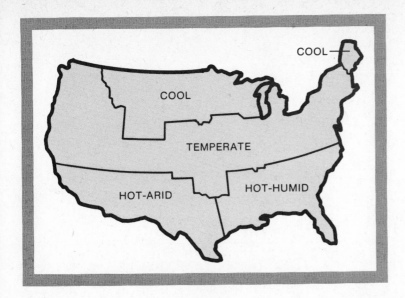

amounts of precipitation are common traits of the temperate regions also. Intermittent periods of clear sunny days are often followed by extended periods of cloudy overcast days.

HOT-ARID REGIONS: Hot-arid regions are characterized by clear sky, dry atmosphere, extended periods of overheating, and a large diurnal temperature range. Wind direction is generally along an E-W axis with variations between day and evening.

HOT-HUMID REGIONS: High temperature and consistent vapor pressure are characteristic of hot-humid regions. Wind velocities and direction vary throughout the year and throughout the day as well. Wind velocities of up to 120 mph may accompany hurricanes which can usually be expected from a S-SE direction.

CLIMATE AND SOLAR HOME DESIGN

There are four elements of climate which are particularly important for solar home and solar system design. These are solar radiation, air temperature, humidity and air movement. Regardless of the area where you intend to build, it is important that there be a careful analysis of these elements as they relate to your specific site. Solar radiation reaches a given site in three ways: direct, diffuse and reflected. The radiation that finally reaches the earth's surface is called insolation and it arrives in either direct (parallel) rays or diffuse (non-directional) rays. The solar radiation reaching a building includes these direct and diffuse rays and also radiation reflected from adjacent ground or building surfaces. It is these three sources of solar radiation which may be used to heat and cool a dwelling.

An analysis of selected temperature differences in the area where the home is to be built will establish the necessary heating and cooling load of the building. Any solar engineer will want to plot the temperatures (through an established method called degree days) for the area of construction along with the insolation to establish what is called a "figure of merit" which indicates, in general terms, a relative feasibility for a solar system in that particular area.

Humidity of the air can be described in two ways: absolute humidity and relative humidity. Relative humidity is the more useful of the two quantities because it offers a direct indication of evaporation potential. This information becomes crucial for maintaining indoor occupant comfort and also for designing solar cooling systems which utilize evaporative cooling techniques.

The total heat content of air during conditions of high humidity and high temperature is substantially larger than during periods of low humidity and high temperatures. Cooling, of course, is the most effective effort of restoring comfort. This can be accomplished by natural ventilation or mechanical cooling. Solar cooling should definitely be considered if conditions of high humidity and high temperature are prevalent during a large portion of the year.

The fourth factor is air movement. Air movement is measured in terms of wind velocity and direction. The wind in a particular location may be useful for natural ventilation at certain times of the year, and detrimental to the thermal performance of a building or solar collector at other times.

Summer winds, if properly directed by the natural topography or site design and captured by the dwelling, can substantially reduce or eliminate the need for a mechanical cooling system.

Winter winds, on the other hand, can be quite detrimental to the thermal behavior of the dwelling or solar system. Cold winds will increase the surface conductance of the dwelling's exterior wall, thereby increasing its heat loss. A home's heat loss can be reduced by the careful selection and combination of building materials and attention to the shape and position of the dwelling on the site. Location of certain landscaping factors can also help reduce heat load.

Wind can also become an issue in snowfall areas where drifting is likely or in high wind areas where additional structural support of large collectors may be necessary. A design that will minimize drifting snow, particularly on or around the collector is an important consideration as well.

Selecting Your Site

The building site is an extremely important solar design consideration. Site planning for the utilization of solar energy is concerned with two major issues: 1) access to the sun and 2) location of the building on the site to reduce its energy requirements. And too, each climatic region discussed previously has its own distinctive characteristics and conditions that influence site planning and dwelling design for solar energy utilization and energy conservation.

In the cool region, for example, maximum exposure of the dwelling and solar collector to the sun is the primary objective of the site planner. Sites with south-

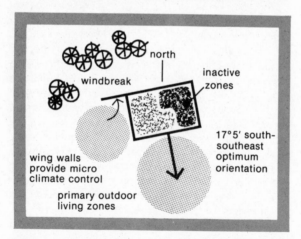

facing slopes are advantageous because they offer the most exposure to solar radiation. Outdoor living areas should be located on the south sides of buildings to take advantage of the sun's heat. Exterior walls and fences can be used to create sun pockets and to provide protection from chilling winter winds.

Locating the dwelling on the leeward side of a hill or in an area protected from prevailing cold northwest winter winds will conserve energy. Evergreen vegetation, earth mounds (berms) and windowless insulated walls can also be used to protect the north and northwest exterior walls of buildings from cold winter winds.

One other consideration is the building of a structure right into a hillside, or partially covering it with earth and natural planting for natural insulation.

In the temperate region, it is vital to assure maximum exposure of the solar collectors during the

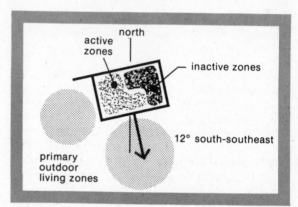

spring, fall and winter months in particular. To do so, the collector should be located on the middle to upper portion of any slope and should be oriented within an arc of 10° either side of south. The primary outdoor living areas should be on the southwest side of the dwelling for protection from north or northwest winds. Only deciduous vegetation should be used on the south side of the home since this provides summer shade and allows for the penetration of winter sun.

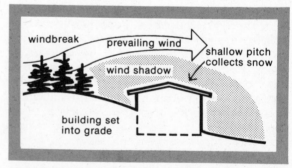

The cooling impact of winter winds can be reduced by using existing or added land forms or vegetation on the north or northwest sides of the dwelling. The structure itself can be designed with steeply pitched roofs on the windward side, thus deflecting the wind and reducing the roof area affected by the winds. Blank walls, garages, or storage areas can be placed on the north side. To keep cold winter winds out of the dwelling, north entrances should be protected with earth mounds, evergreen vegetation, walls or fences.

Outdoor areas used during warm weather should be designed and oriented to take advantage of the prevailing southwest summer breezes.

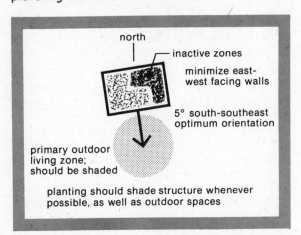

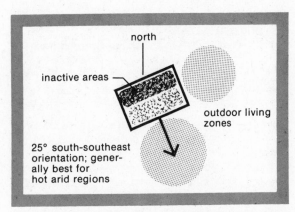

In hot-humid regions where the heating requirement is small, solar collectors (for heating systems only) require maximum exposure to solar radiation primarily during the winter months. During the remainder of the year, air movement in and through the site and shading are the most important site considerations. However, for solar cooling or domestic water heating, year 'round solar collector exposure will be required. Collector orientation within an arc 10° either side of south is sufficient for efficient solar collection.

In the hot-arid regions, the objectives of siting, orientation and site planning are to maximize the duration of solar radiation exposure on the collector and to provide shade for outdoor areas used in late morning or afternoon. To accomplish these objectives, the collector should be oriented south-southwest and the outdoor living areas should be located to the southeast of the dwelling in order to utilize early morning sun and take advantage of shade provided by the structure in the afternoon.

Indoor and outdoor activity areas should take maximum advantage of cooling breezes by increasing the local humidity level and lowering the temperature. This may be accomplished by locating the home on the leeward side of a lake, stream or other bodies of water if possible. Also, lower hillside sites will benefit from cooler natural air movement during early evening and warm air movement during the early morning.

Exterior wall openings should face south but should be shaded either by roof overhangs or by deciduous trees in order to limit excessive solar radiation into the dwelling. The size of windows on the east and west sides should be minimized in order to reduce radiation heat gain in early mornings and late afternoons.

GENERAL CRITERIA FOR SITE SELECTION

- Geography of the surrounding area:
  the daily and seasonal patterns of sun and windflow, as well as earthform impediments and low areas where cold could settle
- Topography of the site:
  steepness of the slope in terms of building economy, slopes beneficial to energy conservation or solar utilization
- Orientation of slopes on the site:
  slope direction to determine most beneficial solar exposure
- Geology underlying the site:
  depth and type of rock on site and unbuildable areas
- Existing soil potential/constraints:
  can the soil support a structure and can it support vegetation?
- Existing vegetation:
  size, variety and location of vegetation which would impair or enhance solar collection or energy conservation, or which would be disturbed by building

- Climatically protected areas on site:
  areas protected at certain times of day or year, by topography, vegetation
- Climatically exposed location:
  areas exposed to sun or wind, in summer, winter, all year long
- Solar radiation patterns on the site:
  daily, monthly, including impediments such as vegetation, etc.
- Wind patterns on the site:
  same considerations as above
- Precipitation patterns on the site:
  fog movement, collection and propensity patterns, snow drifting and collection, frost pockets
- Temperature patterns on the site:
  daily, monthly, seasonal as well as warm and cold pockets
- Water/air drainage patterns:
  seasonal and daily water and air flow patterns and natural impediments to these patterns

The Structure as a Planning Variable

We have reviewed how climatic conditions affect the utilization of solar energy in a dwelling, and we have seen how the negative aspects of these climatic conditions can be minimized by taking full advantage of the siting situation. The structure of the home itself, of course, is also a major controlling variable.

A building gains or loses heat by conduction, convection, evaporation, radiation, internal heat sources and mechanical systems.

The thermal balance of a building is maintained if the heat lost and gained from the above sources equals zero. A dwelling designer and builder share the responsibility of properly selecting building materials, determining building size, volume and orientation, and sizing and orienting windows, doors, overhangs, and other thermal controls to assure occupant comfort. Each of these considerations will influence the magnitude of one or several of a building's heat exchange processes. The trade-offs between heat exchange factors will most likely be climate, cost and construction practices. For example, since mechanical equipment is expensive, it may be appropriate to reduce heat loss and gain by structural methods. By working with instead of against climatic impact, a reduction of a building's need for mechanical equipment may be realized. This, in turn, may result in a reduction in the amount of fossil fuel or the size of a solar system required to adequately heat and/or cool a home.

Thermal controls are devices (furnace) or methods (dwelling orientation to capture summer winds) for moderating the extremes of outdoor climate to bring the interior within the narrow ranges of temperature and humidity that support human comfort. There are numerous ways to analyze thermal control, but for the purposes of this book, it is probably better to review them in terms of the climatic variable they are intended to control . . . temperature, sun and wind. The emphasis in this book will be on control strategies which do not require a conventional energy input for their operation.

Temperature can be regulated by either thermal retention or thermal regulation. An example of the former would be the use of resistive insulation and of the latter would be internal zoning or strategical landscaping.

Strategies for sun control can also be organized into two categories: solar exposure and light regulation. Solar exposure methods moderate the exposure of the building and adjacent site to solar radiation. Light regulation, on the other hand, regulates the amount of sunlight reaching the interior of the dwelling. Insulation and solar collectors are also considered solar exposure strategies, while window sizing and placement and shading fall into the light regulation strategies.

Wind regulation techniques range from designing the building itself as a wind-controlling form through aerodynamic massing, to placing natural or man-made elements to regulate wind direction and force.

The challenge that confronts the designer is to select that combination of thermal controls which moderates the climate to expected levels of occupant comfort at a justifiable cost in construction and energy consumption.

Right: The Bicentennial Energy-Efficient Home, a cooperative effort of James Molinaro of Molinaro & Son, Inc., builder, and Pennsylvania Power & Light Co., was designed to direct attention toward our current energy picture and encourage prospective new homeowners to plan for the future in the design and construction of new homes. Energy-saving features incorporated in this home include:
- Hi-Re-Li heat pump with supplemental electric resistance heat
- 1″ extruded polystyrene T&G insulating exterior sheathing
- Friction fit batt insulation (R-13) in sidewalls
- Foil back plaster board vapor barrier
- 6½″ batt insulation (R-22) in ceiling of 2nd floor
- 6″ batt insulation (R-19) in basement ceiling

- 6″ batt insulation (R-19) in box ends between 1st and 2nd floor
- Foamed-in-place polymeric foam sealant around all windows and exterior doorways
- Insulated glass wood sash windows with separate storm sash
- Insulated exterior doors
- Air-to-air heat exchanger for make-up air
- Insulated heating ducts on air-handling system in unheated space
- Low wattage 80 gal. water heater with solar assist
- Insulated hot water pipes in unheated space
- Adequate attic ventilation, ridge, gable and eave vents
- Outside air ducted directly into fireplace combustion area
- Tempered glass fireplace screen

Solar System Components

Several characteristic properties apply to all solar heating/cooling and domestic hot water systems, whether they are simple or relatively complex. Any solar system consists of three generic components: collector, storage and distribution; and may include three additional components: transport, auxiliary energy system and controls. These components may vary widely in design and function. They may, in fact, be one and the same element (a masonry wall can be seen as a collector, although a relatively inefficient one, which stores and then radiates or "distributes" heat directly to the building interior). They may also be arranged in numerous combinations dependent on function, component compatibility, climatic conditions, required performance, and architectural requirements.

Solar energy, also known as solar radiation, reaches the Earth's surface in two ways: by direct (parallel) rays; and by diffuse (non-parallel) sky radiation, reflected from clouds and atmospheric dust. The solar energy reaching the surfaces of buildings includes not only direct and diffuse rays but also radiation reflected from adjacent ground or building surfaces. The relative proportion of total radiation from these sources varies widely in each climate, from hot-dry climates where clear skies enable a large percentage of direct radiation to reach a building, to temperate and humid climates where up to 40 percent of the total radiation received may be diffuse, to northern climates where snow reflection from the low winter sun may result in a greater amount of incident radiation than in warmer but cloudier climates. As a result of these differences in the amount and type of radiation reaching a building, as well as in climate, time of year and type of use — space heating, cooling or year-round domestic water heating — the need for and the design of solar system components will vary in each locale. Recognition of these differences is important for the proper design and/or selection of solar system components.

In the following pages, the various methods of collecting, storing and distributing solar energy will be discussed and illustrated. The individual solar components will then be assembled into solar heating/cooling and domestic hot water systems and the process by which solar radiation provides heating and cooling discussed. A simplified diagram of a solar heating system is presented below.

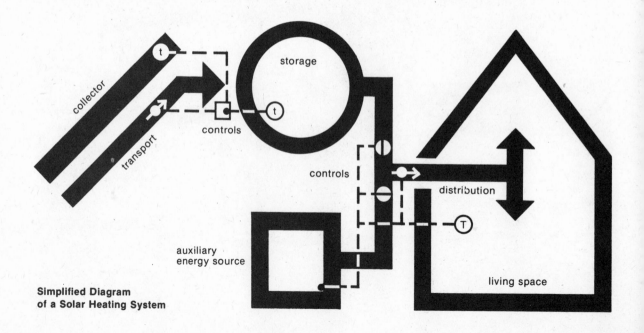

**Simplified Diagram
of a Solar Heating System**

1 Solar Collectors

There are numerous concepts for the collection of solar radiation. These concepts range from the most simple — a window — to those that are quite complex and require advanced technology for their development — a solar cell for instance. Historically, solar collectors have been classified as either focusing or non-focusing.

A non-focusing collector is one in which the absorber surface is essentially flat and where the absorber area is equal to the aperture for incident radiation. A focusing collector, however, is one in which the absorber area is smaller than the aperture for incident radiation and consequently there is a concentration of energy onto the absorber surface. Numerous solar collectors have been developed which illustrate each of these concepts. Several recent collector designs have been developed which do not fit into either category, creating considerable confusion in collector classification. Consequently, the following description of solar collector concepts is not separated exclusively into focusing or non-focusing collectors.

The brief concept descriptions offer a general classification of solar collectors. The reason for collector classification is to introduce you to a range of solar collector concepts which may be used singly or together for the capture of solar radiation and to indicate in general terms their present applicability.

FLAT-PLATE COLLECTORS
Of the many solar heat collection concepts presently being developed, the relatively simple flat-plate collector has found the widest application. Its low fabrication, installation, and maintenance cost as compared to higher temperature heat collection shapes has been the primary reason for its widespread use. Additionally, flat-plate collectors can be easily incorporated into a building shape provided the tilt and orientation are properly calculated.

Flat-plate collectors utilize direct as well as diffuse solar radiation. Temperatures to 250°F (121°C) can be attained by carefully designed flat-plate collectors. This is well above the moderate temperatures needed for space heating, cooling and domestic water heating.

A flat-plate collector generally consists of an absorbing plate, often metallic, which may be flat, corrugated, or grooved; painted black to increase absorption of the sun's heat; insulated on its backside to minimize heat loss from the plate; and covered with a transparent cover sheet to trap heat within the collector and reduce convective cooling of the absorber. The captured solar heat is removed from the absorber by means of a working fluid, generally air or treated water, which is heated as it passes through or near the absorbing plate. The heated working fluid is transported to points of use or to storage.

Three types of flat-plate collectors will be discussed. There are innumerable variants, but the following serve as an introductory classification. In selecting any particular collector, one should consider thermal efficiency, the total area and orientation required, durability of materials, and initial operating cost.

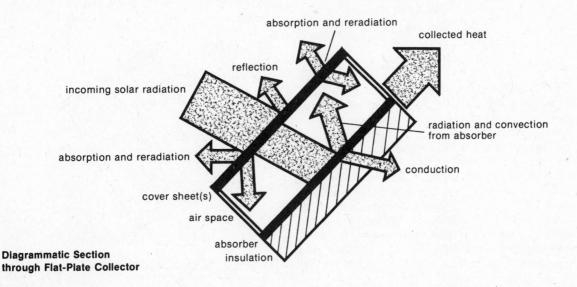

absorption and reradiation

collected heat

reflection

incoming solar radiation

radiation and convection
from absorber

absorption and reradiation

conduction

cover sheet(s)

air space

absorber
insulation

**Diagrammatic Section
through Flat-Plate Collector**

OPEN WATER COLLECTOR

At present, collectors which are factory-produced and shipped to the building site are relatively high in cost due in part to the small volume manufactured. Collectors built from commonly available materials and fabricated on the site are less expensive. Their thermal efficiency, however, may be lower than factory-produced units. An open water collector is representative of on-site fabricated collectors which use corrugated metal roofing panels painted black and covered with a transparent cover sheet. The panels thus provide open troughs in the corrugations for trickling water to be fed from a supply at the top of the roof to a collection gutter at the base, where it is then transported to storage. Heat losses that occur by evaporation in open systems are reduced in some designs by the nesting of two corrugated sheets with a small enough passage in between them to force the water into contact with the top sheet. Open water collectors should be carefully evaluated before use in cold climates, to determine the extent of condensation and corresponding loss of efficiency.

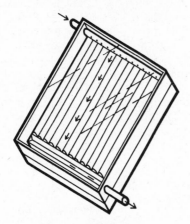

Open Water Collector

AIR-COOLED COLLECTOR

Collectors that employ air (or gas) as the transport medium between collector and storage have been developed. Low maintenance and relative freedom from the freezing problems experienced with liquid-cooled collectors are two of the chief advantages of air collectors. In addition, the heated air can be passed directly into the dwelling space or into the storage component. Disadvantages are the inefficient transfer of heat from air to domestic hot water and the relatively large duct sizes and electrical power required for air transport between collector and storage. Although few air collectors are now readily available from manufacturers, in contrast to the more than one dozen sources of liquid-cooled collectors, it is predicted that air-cooled collectors will soon become more widely used.

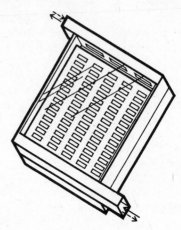

Air-Cooled Collector

LIQUID-COOLED COLLECTOR

Most collectors developed since the time of the M.I.T. experimental houses have used water or an anti-freeze solution as the transport medium. The liquid is heated as it passes through the absorber plate of the collector and then is pumped to a storage tank, transferring its heat to the storage medium.

The prevention of freezing, corrosion and leaks has been the major problem that has plagued liquid-cooled systems. This is accomplished by using oil or water treated with corrosion inhibitors as the transport medium or by designing the collectors to drain into storage during periods of non-collection.

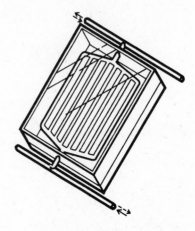

Liquid-Cooled Collector

INCREASING PERFORMANCE

Flat-plate collectors are frequently mounted on the ground or on a building in a fixed position at prescribed angles of solar exposure — which vary according to the geographic location, collector type, and use of absorbed solar heat. The fixed mounting has advantages of structural security and design integration, but must be oriented within prescribed limits to receive a level of solar radiation commensurate to the capital investment involved in installing a solar energy system. For space heating, a tilt of latitude plus 15° and an orientation true south to southwest (afternoon air temperatures are higher and thus bias an orientation west of true south) are considered optimal, based on existing solar experiments in operation at this time. For combined space heating and cooling, collector orientation remains the same while collector tilt is changed to latitude plus 5°. Variations are generally tolerable only within 10° to 15° of these optima. There have been several proposals to improve the annual thermal performance of flat-plate collectors. One concept involves the use of adjustable/sun-tracking flat-plate collectors that are continually or periodically adjusted, while another one uses reflecting panels to increase total thermal yield.

Mounting a flat-plate collector on adjustable or sun-tracking mechanisms can improve the annual thermal performance of the collector by as much as 70 percent. A collector can be tilted from the annual optimum (described as latitude plus 15° for space heating) to more closely approximate the seasonal optima. For instance, season optimum for space heating will vary 47° from December to June. A monthly adjustment can be made manually, provided there is access to the collectors, which can yield at least a 10 percent improvement over a fixed position. Adjustable tilt mechanisms can usually be accommodated most easily on ground or flat roof installations.

A collector can also be rotated on one axis to follow the daily path of the sun's orientation from east to west. This will require an automatic tracking mechanism, unless a method for manual adjustment during the day is provided. With an automatic sun-tracking mechanism, annual energy increases of 40 percent could be expected in areas where the predominant source of thermal energy is direct radiation.

The tilt and orientation of a flat-plate collector can be maintained in any optimum position throughout the day and year by use of a heliostatic mount. A heliostatic mechanism maintains the collector at a perpendicular (normal) exposure to direct solar radiation. Although the control mechanism and structural mounting are complicated, a sun-tracking flat-plate collector could collect 70 percent more solar radiation than the same flat-plate collector in a fixed optimum position. The cost of such a sophisticated tracking solar collector should be compared with the cost of a larger fixed collector which would deliver the same energy output.

Another concept for increasing the thermal yield per unit of flat-plate collector area is the use of panels which reflect additional solar radiation onto the collector, thereby increasing its thermal performance. The same panels, if so designed, can also be used to cover and insulate the collector during non-collection periods, at night, and on cloudy days. The panels may be operated by sun-sensitive automatic controls or manually. In either case, questions of maintenance and operation should be considered, particularly in areas that experience snow or ice.

There are also collection arrangements that reflect incoming solar radiation several times onto a focused or concentrated area, thereby using an optical gain to increase the unit collection of the absorbing surface. As a result, the area of collector absorbing surface needed is reduced. There are numerous concepts based on such principles. One patented idea that has found application is a flat-plate collector that is located in a roof shed where an adjustable reflecting panel directs the radiation to an otherwise weather-protected collector.

Flat-Plate Collector with Adjustable Tilt

CONCENTRATING COLLECTORS

Concentrating collectors use curved or multiple point-target reflectors to increase radiation on a small target area for either a tube or point absorber. Presently, concentrating collectors are greater in cost than flat-plate collectors, with added problems of reflective surface maintenance. For use in sunny climates, however, they promise more than double the temperature generated by flat-plate collectors. Concentrating collectors are best suited for areas with clear skies where a major portion of solar radiation is received in direct rays. Their inability to function on cloudy or overcast days is a significant disadvantage of concentrating collectors as compared to flat-plate collectors. However, they may find a particularly viable role as a collector for solar cooling systems, but at the present time they require more development than flat-plate collectors.

Linear concentrating collector ● A reflector curved in one direction that focuses radiation on a pipe or tube absorber is a linear concentrating collector. Heat is removed from the absorber by a working fluid circulating through the pipe and transported to point of use or storage. The absorber is generally covered with a transparent surface to reduce convective or radiative heat losses. The working fluid should have a boiling point above the expected operating temperatures of the collector and also be resistant to freezing.

Linear concentrating collectors can be designed with the long axis horizontal (in an east-west direction) or at an optimum tilt (in a north-south direction). Since the change in the sun's altitude during the day is less than the change in its compass direction, horizontal linear concentrating collectors can be designed for manual adjustment every few days, to track the average path of the sun as it changes from season to season. Tilted linear concentrators with the long axis north-south must track the sun throughout the day, as must horizontal linear concentrators designed for precise focusing.

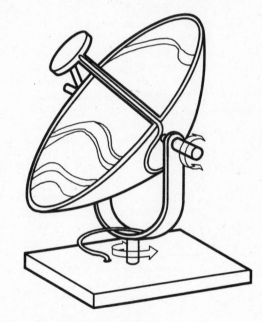

Circular Concentrating Collector

Circular concentrating collector ● A reflector in the shape of a dish or hemisphere that is used to focus solar radiation on a point target area is a circular concentrating collector. An absorber located at the focal point absorbs solar produced heat where a working fluid then transports it away from the collector. The collector may be fixed and the target area (absorber) movable to accommodate the daily path of the sun as the focus point changes with the direction of incoming rays. More commonly, the entire reflector and absorber assembly is made to follow the sun. Several experimental designs have been constructed utilizing this principle. The high temperatures achievable by such collector mountings may eventually justify their use despite problems of operation, durability, design integration and structural mounting, if they permit economies in the design of the total system or the building itself.

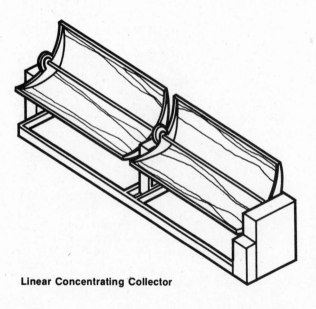

Linear Concentrating Collector

PASSIVE COLLECTORS

The collector concepts discussed to this point have been relatively independent elements which can be organized and operated quite apart from the building itself. That is, the relationship of the collector to the building it serves can vary significantly without a major alteration of system performance. There is a point when it is advantageous in terms of cost, building design and system operation that the collector, storage and building be physically integrated.

There are numerous collector concepts where the relationship of the collector to the building is direct and the alteration of collector design will modify in varying degrees the building's design. These concepts use the entire building or various elements of the building (walls, roof, openings) as solar components. As such, the collector and building are one and the same element and therefore cannot be separated from each other. This type of collection method has come to be known as inherent or passive solar collection.

The integrated nature of passive solar collectors permits the design of a variety of collector concepts. The imagination of the designer and builder is the only limiting factor. Passive collectors can be as simple as a window or greenhouse.

Three general passive collector concepts will be discussed. There are innumerable other concepts, but the following will serve as an introduction to inherent or passive solar collectors.

Below: A passive solar home — architect Lee Porter Butler has this to say about his designs: "My method is not a system for collecting and storing heat. It's a system that optimizes the thermal inertia of mass, plus orientation and insulation techniques that are well known. The radical, revolutionary aspect of this is that you can get rid of the need for any kind of heating system." For more on this home designed by Mr. Butler, please refer to Chapter 3.

Incidental heat traps ● All collectors are "heat traps" in that they capture heat from direct solar or diffuse sky radiation or from adjacent ground or building surface reflection. There are numerous building components including windows, roof monitors, and greenhouses which are not normally considered solar collectors but can be used as such along with their other principle function. Because they can serve various purposes — visibility, ventilation, natural illumination — as well as incidental heat collection, they become particularly valuable building components in integrated solar dwelling design. Additionally, they illustrate that energy conservation does not preclude the use of windows, glass structures, or skylights if they are designed to maximize direct solar heat gain.

The use of south-facing windows to increase heat gain directly into a building is well known. The concept

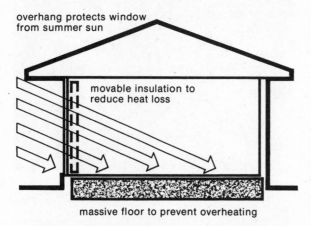

overhang protects window from summer sun

movable insulation to reduce heat loss

massive floor to prevent overheating

South-Facing Window as a Solar Collector

was used in the United States in popular house plans that were often referred to in the 1940's and 1950's as "solar homes." The south wall orientation is considered ideal compared to east and west walls because in northern climates shading of a south window to prevent summer overheating is easily accomplished by an overhang calculated to equinox sun angles. In such arrangements, careful attention must be given to insulating the window at night, preferably from the interior, to reduce heat loss. With an interior insulating drapery, or even better, a shutter which is much more air-tight, windows do indeed function as effective heat traps and have been shown to be able to provide a sizable percent of the annual heat requirements of a building (estimates vary from 25 to 60 percent, depending on climate and use).

Windows used as solar collectors have the drawback of overheating the space they serve. In order to reduce the overheating effect, masonry surfaces such as concrete, brick, tile, or stone on the floor or on the walls can be used for their heat storage capacity, absorbing the heat during the day and radiating it subsequently for several hours or more. The storage effect of a particular floor or wall can be calculated as a function of the specific heat (specific heat is the quantity of BTU's which can be stored per pound per degree F) of the masonry, its volume and weight, and the expected temperature differences it will experience throughout the day. Too great a storage effect in the exposed room surfaces can have a negative effect on occupant comfort or fuel consumption if the morning "reheat" time of the materials is too long. However, properly designed, the thermal mass of construction materials can play a significant role in an integrated solar dwelling design.

The use of a greenhouse as an incidental heat trap is a further elaboration of the solar window concept. It exposes more glass area to solar radiation than a window but with greater heat loss if no provision is made for insulation.

Again, storage such as masonry surfaces or a rock pile under the greenhouse may be necessary to avoid overheating the space. One advantage of a greenhouse as a solar collector is that it can be closed off from the rest of the house on sunless days, thereby reducing the net heated area of the dwelling. However, during some periods of cooler weather it can be used as a supplementary space of the house — an atrium, sun room, or a garden room.

Another passive collector concept valuable because of its versatility is a roof monitor. A roof monitor is a cupola, skylight, or clerestory shed arrangement that is designed to control heat gain, natural light, and/or ventilation. Roof monitors by themselves are poor devices for gaining usable heat: first, because the heat enters at the high point of the building; and second, because the roof exposure gains so much solar radiation in the summer that shading or insulating arrangements are required. Also, because of the low winter sun angle, the sun entering through the roof monitor would probably not reach the floor or lower wall surfaces to offer any storage effect. However, if a return air register is located at a high point in the space, the trapped heat can be recirculated. This technique will prevent a temperature stratification within a building by continually returning solar heat gained through the building and roof monitor to the lower occupied spaces.

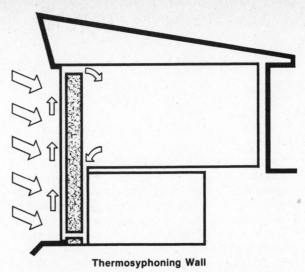

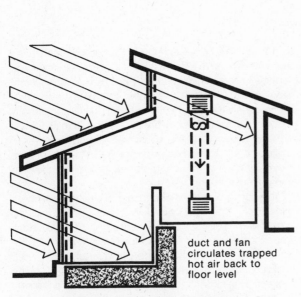

Roof Monitor as a Solar Collector

Roof monitors are a particular interest because they are excellent sources of natural lighting and can be used in summer months to augment natural cooling through the "thermal chimney" effect. (The thermal chimney effect refers to the natural rise of hot air in a building which is vented to the exterior, thereby causing a continuous circulation of air for ventilating purposes.) Roof monitors, designed with proper insulating and ventilating controls, can be used to great advantage in a low energy approach to design.

Thermosyphoning walls/roof ● Another set of passive collector concepts makes use of heat that is built up within a wall or roof structure by "syphoning" or drawing it off and supplying it to a room or storage element. "Thermosyphoning," a term traditionally applied to mechanical systems that use the natural rise of heated gases or liquids for heat transport, is the primary method for moving captured heat to point of use or storage. To avoid overheating in the summer, the space where heat builds is vented out.

One thermosyphoning concept which has found application in traditional as well as solar designs is the use of solar heat trapped in air spaces in walls and roofs. When the trapped air temperature exceeds the temperature of the internal building space, it can be drawn off by direct venting or forced air ducts.

This method of solar heat collection is marginal at best due to the small amount of heat collected, problems in the control of temperature differences between the inside of the wall or roof structure and the occupied space, and the large ducts and electrically powered fans needed to move any sizable volume of heated air. However, the concept of using the solar

heat that is built up within a building's walls or roof deserves consideration as a multi-purpose solution to annual climate conditions.

A more effective variant of the preceding concept is one where the external surface or internal and external wall or roof surface are transparent. The heated air trapped between the building surfaces can be used because it will usually be hotter than the temperature of the occupied space. This concept is not as effective a formal solar heat collector (i.e., flat-plate collector) but it has the advantages of admitting natural light, and providing better insulation than a plate glass window.

In a further elaboration, the building envelope can also be used as storage. A glass wall is placed over an absorbing material — generally masonry — which is painted dark and serves as heat storage for a time-lag capacity that has been previously calculated. The air

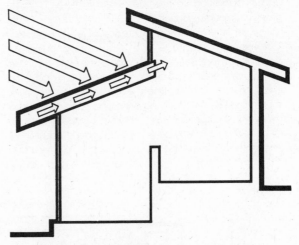

Thermosyphoning Roof as a Solar Collector

space between the glass wall and the absorbing material is vented to the interior at the top of the wall or ducted to rock storage elsewhere in the building. A cold air return must be located at the bottom of the collector so that a thermosyphoning arrangement can be used to facilitate air circulation. The morning reheat time of large storage media becomes a design calculation that is particularly critical.

Solar ponds ● A solar pond is a particularly interesting passive collector concept because it can provide for both heating and cooling. Also, a solar pond may be integral with the building structure — on the roof for example — or entirely separated from the building on adjacent ground. For either situation, control must be maintained over the heating and cooling processes for efficient operation. This can be accomplished by the use of movable insulating panels to expose or conceal the pond, by filling and draining the pond according to the heating and cooling demand or by covering the pond with a transparent roof structure.

Although solar ponds at present have found limited application confined to the southwest, they have also been proposed for use in northern climates. Solar ponds are particularly appropriate to climates where the need for cooling is the principle design condition and where summer night temperatures are substantially lower than daytime temperatures. The combination of these climatic conditions, found usually in hot-arid regions, permits the ponds and elements of the building (exterior walls) to be cooled by natural radiation to the night sky and to effect a time-lag of temperature through the building envelope, in ideal cases up to eight or more hours. This permits numerous design concepts to aid natural cooling by radiation and evaporation. In hot-humid climates, high vapor pressure, cloudiness and a small diurnal temperature range limits the cooling efficiency of solar ponds.

The major advantage of a solar roof pond is that it does not dictate building orientation or exposure, and as long as there are no barriers between sun and pond it will provide a completely even heating or cooling source over the entire living area of a building. Several dwellings incorporating a solar roof pond have been designed and constructed. The Atascadero house is an example of a dwelling concept using a roof pond. In this patented design, water containers are disposed on a flat roof and covered with insulating panels. During summer days the panels are closed. At night the panels are removed to lose heat to the cooler night sky. Suitably cooled, the water containers then draw heat during the day from the building interior (usually through a metal deck roof). In winter, the process is reversed, exposing the containers during the day to collect and store solar thermal radiation and then covered by the insulating panels at night, providing heat to the interior.

The same process is applicable in northern climates. However, the roof pond is now covered by a transparent roof structure oriented to receive maximum incident solar radiation. Heat is trapped in the attic space thus warming the pond. Insulating panels cover the transparent surface during periods of no collection in order to reduce heat loss. The transparent surface could also be removed during the hot, humid summer months to increase the evaporative and convective cooling effect.

An alternative to water containers covered with insulating panels or enclosed in an attic space is a roof pond with circulating water between the roof and a storage tank located below or within the occupied space. With this system the need for movable insulating panels is eliminated. During the heating cycle, circulation would take place during the daytime. The water, heated by the sun, would be stored or distributed depending on the dwelling's heating requirement. To prevent cooling by evaporation the pond must be covered by a transparent surface — glass or plastic which can float on the surface. During the cooling cycle, the circulation of water would take place only during the night. The cooled water would be stored to draw heat during the day. The transparent cover sheet would no longer be needed for efficient cooling operation, since cooling would be achieved by night sky radiation, convection and evaporation.

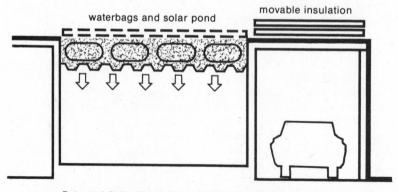

waterbags and solar pond   movable insulation

Patented Solar Pond Concept Used in Atascadero House

2 Solar Heat Storage

The intermittent availability of solar radiation requires that heat be stored during times of favorable collection for later use for such purposes as space heating, cooling and domestic water heating. In many cases the use of several storage methods or the storage of heat at different temperatures has been shown to be advantageous for supplying the heating and cooling demand of specific buildings. The type, cost, operation, and required size of the solar storage component will be determined by the method of solar collection, the dwelling's heating and cooling requirement and the heat transfer efficiency to and from the storage unit.

The following heat storage concepts are classified in broad categories. Each storage method falls within two generic methods: sensible heat storage and latent heat storage. The technical feasibility of each method has been demonstrated either in actual use or in experimental testing.

SENSIBLE HEAT STORAGE

This concept means that solar heat may be stored by raising the temperature of inert substances such as rocks, water, masonry, or adobe.

Room air and/or exposed surfaces ● The solar radiation received from south facing windows or transparent panels increases the temperature of room air and surfaces exposed to the sun's rays. As such, the room's air and exposed surfaces (walls, floors, etc.) are the solar storage component for a window wall, greenhouse or transparent panel which collects the radiation. For most situations the storage capacity of the air and surfaces will not be sufficient for long periods of heating demand. Additionally, in the process of "charging" the storage, the space may become overheated and possibly extremely uncomfortable for the occupants.

To use effectively the radiation stored in the air and room surfaces, careful attention must be given to minimizing the loss of heat at night or when collection is not occurring. Insulated drapes, shutters, and other such devices are necessary to reduce heat loss and increase the use of trapped heat. The room size, the window placement, the material composition, volume and weight, and the expected temperature difference will also determine the performance of the solar storage.

A more direct application of this storage concept involves the placement of a glass or transparent wall

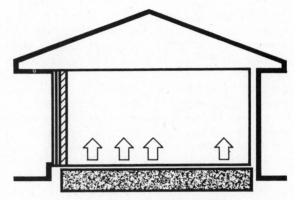

Heat Storage in Massive Floor

over an exposed masonry surface such as concrete. The exposed surface which serves as the heat store is painted a dark color and located directly behind the transparent surface. The thermal conductivity and specific heat of the wall material and the expected temperature range will determine the volume of the wall. The reradiation time-lag must be accurately calculated to assure proper heating of the space. It is sometimes necessary to place insulation on the room side of the storage wall to avoid overheating the space. The exposed masonry storage method is usually used in conjunction with interior and exterior vents to control the heat distribution to the space or to another storage system.

A variation of exposing a masonry surface to solar radiation is to expose containers filled with water. The exposed water containers may be placed on the roof or used as interior or exterior walls. The previously described Atascadero house which uses plastic bags filled with water placed on the roof is an example of this method of storage. Again, careful calculations are required to properly size the storage capacity. Also, a similar means of thermal control is necessary to assure a proper lag-time and to avoid overheating.

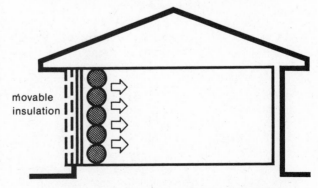

movable insulation

Heat Storage in Water Tanks behind Wall Collector

Rock storage • A common method of heat storage, most often associated with air-cooled flat-plate collectors, is rock storage. Pebble beds or rock piles contained in an insulated storage unit have sufficient heat capacity to provide heat for extended sunless periods. The rock storage is heated as air from the collector is forced through the rock container by a blower. Rock storage will require approximately 2½ times the volume of water storage, assuming the same temperature range. For example, a rock pile with a void space of one-third of the total volume can store approximately 23 BTU/cu.ft./degree F, while water can store 62.5 BTU/cu.ft./degree F.

A convenient rock size for storing solar heat is about 2 inches in diameter. A decrease in pebble size increases the air flow resistance through the storage and may affect blower and duct size and distribution efficiency. Unlike the previous storage methods, rock storage does not have to be in close proximity to the collector. However, as the distance increases, the heat transfer losses between the heated air and the rocks

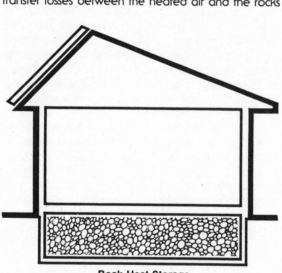

Rock Heat Storage

also increases and larger air ducts and more electrical power are generally required for moving air between the collector, storage, and heated spaces.

Water storage • Water has the highest heat capacity per pound of any ordinary material. It is also very inexpensive and therefore is an attractive storage and heat-transfer medium. However, it does require a large storage tank which may be expensive. The storage tank is usually insulated to reduce conductive heat losses. Also, it is sometimes practical to compartmentalize the storage tank to control temperature

gradients (different temperatures within storage tank) and to maintain an efficient heat transfer. Potential disadvantages of water storage include leakage, corrosion and freezing.

Heat is generally transferred to and from storage by a working fluid circulated by an electric pump. The heated working fluid itself may be placed in storage or its heat transferred to the storage tank by a heat exchanger. The process of heat transfer to water is more efficient than to rock and therefore less surface for the heat exchanger is required. With water storage, proximity of storage to the collector is not as critical as with rock storage. Also, compared to rock storage, water occupies a comparatively small volume.

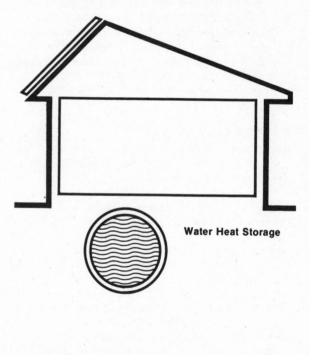

Water Heat Storage

LATENT HEAT STORAGE
The use of the heat of fusion or heat of vaporization associated with changes of state or with chemical reactions offers the possibility of storing a great deal of heat in a small volume. (Heat of fusion is the latent heat involved in changing between the solid and liquid states while heat of vaporization is the latent heat involved in changing between liquid and vapor states.) Although numerous physical/chemical processes have been investigated and offer numerous advantages compared with sensible heat storage, there is not at the present time a completely reliable storage method using latent heat.

3 Thermal Energy Distribution

Generally speaking, there are three methods by which thermal energy from storage or collector can be distributed to point of use: gas flow, liquid flow and radiation. Within each category there are several techniques. Some involve mechanical and electrical equipment and processes while others utilize natural convection and radiation. The manner in which solar radiation is collected and stored will determine the means of distribution. If an air-cooled flat-plate collector is used to capture solar radiation and a rock pile is used to store the heat, distribution is usually by air.

GAS FLOW DISTRIBUTION

This type of thermal energy distribution throughout a dwelling can involve both mechanical and natural methods. Natural convection and forced air are two of the more noteworthy methods.

Natural convection ● Natural convection is the circulatory motion of air caused by thermal gradients without the assistance of mechanical devices. An example of convection is the motion of smoke towards room lamps — the hot air generated by the lamp rises because it is less dense, and cooler air moves in.

Natural convection is a useful means of distributing solar thermal energy because it requires no mechanical or electrical input. However, for this same reason, careful attention to design is required to maintain control of convective distribution methods. Placement of solar collectors, storage, interior and exterior walls and openings is important.

The operation cycle of natural convective distribution is quite simple. Heat from the collector or storage is supplied to the habitable space. This process is controlled by the collector or storage design or by wall or floor vents. As the hot air rises to displace cooler air, convection currents similar to those causing winds occur, and the air is distributed through the space. The air is cooled, becomes dense and falls toward the floor, where it is captured by cool air return vents, passed through the collector and storage and once again distributed. The cycle will continue as long as there is a temperature difference between the collector/storage components and the room air. When the convection heating cycle is not desired, the warmer air may be vented to the exterior.

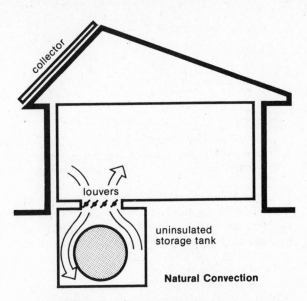

Natural Convection

Forced air ● A forced air system relies on mechanical equipment and electrical energy for the distribution of thermal energy. Design for solar systems is much the same as for conventional forced air systems. However, because solar produced temperatures in storage are often relatively low, distribution ducts and vents must normally be larger than conventional systems. Achieving maximum efficiency in a solar system requires careful attention to air distribution design.

Forced air distribution for solar systems is similar to conventional air distribution. Air from either the collector or storage is blown through ducts to the occupied spaces. The type of solar collector or storage is not the determining factor for selection of a forced air ducted system; the system is adaptable to rock, water or phase change storage components. For rock and containerized phase change storage, air is simply blown through the storage to ducts which supply the dwelling spaces. In the case of water storage, a heat exchanger is required to transfer heat from the liquid to air which is distributed to dwelling spaces.

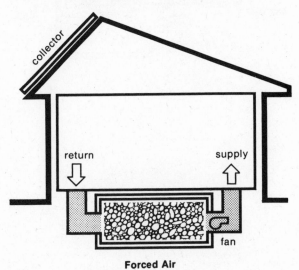

Forced Air

LIQUID FLOW DISTRIBUTION

The distribution of thermal energy using liquids can also be either natural or mechanical. Forced radiation and natural radiation are two principal methods.

Forced radiation ● Forced radiant distribution relies on the transfer of heat to air in the occupied spaces by radiation and convection from circulating hot water through tubes. For cooling, the forced radiant system is generally used with a refrigeration unit which passes chilled water through a fan coil unit located at the point of distribution. A blower forces air through the cooled fan coil unit and into occupied spaces.

The piping for the radiant system may be located in the ceiling, floor, or along the wall in fin tube baseboard units. The only significant alteration required of conventional radiant systems for use by solar systems is the enlargement of the radiating surfaces — larger fin tubes or closer spaced ceiling or floor coils — because of lower temperatures from storage.

Natural radiation ● Natural radiation is the transfer of heat by electromagnetic waves without the assistance of mechanical devices. The radiation properties of the emitting and absorbing surfaces, which are influenced by their temperature, will determine the rate of heat flow between them.

Unlike natural convection, which is dependent on differential air temperatures for distribution, natural radiation is dependent on differential surface temperatures. An example of natural radiation is the sun warming a greenhouse on a cold day. The radiant energy is transferred directly to the greenhouse surfaces and is not significantly affected by the cool temperature of the surrounding air. Natural radiation is particularly useful for collector or storage systems which are directly exposed to the occupied spaces. The captured energy can be emitted by natural radiation directly to the room's surface. The walls, floors, and ceiling which are used to collect and/or store thermal radiation will radiate to a room's other cooler surfaces.

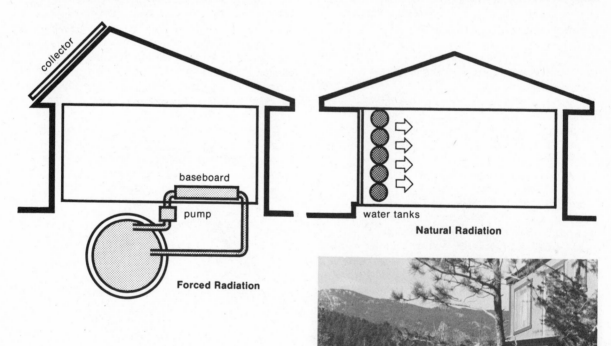

Forced Radiation

collector

baseboard

pump

water tanks

Natural Radiation

Right: An example of natural radiation, greenhouse supplies supplemental heat and humidity to the home. Greenhouse is pre-fabricated by Solar Technology Corp.

Solar Systems

Solar system design, in its many different approaches, assembles the collector, storage, and distribution components into a heating, cooling and/or domestic hot water system. Each component of a solar system (collector, storage, and distribution) may be compatible with a limited number of other solar components or may be compatible with many. For example, a solar collector may be compatible with a specific storage component which in turn may serve one or several types of distribution systems.

To illustrate the compatibility of various solar components and to describe the process of converting solar radiation into thermal energy for heating and cooling, several representative solar systems will be discussed. Each representative solar system will be made up of a collector, storage, and distribution component described previously. A solar system may be diagrammed as illustrated.

The basic function of a solar system is the conversion of solar radiation into usable energy. This is accomplished in general terms in the following manner. Radiation is absorbed by a **collector**, placed in **storage**, with or without the assistance of a **transport** medium, and **distributed** to point of use — an occupied space. The performance of each operation is maintained and monitored by automatic or manual **controls.** An **auxiliary energy system** is usually available for operation, both to supplement the output provided by the solar system and to provide for the total energy demand should the solar system become inoperable at any time.

With this relatively simple process in mind, a more detailed explanation of several solar systems is presented on the following pages, illustrating some of the many variations in solar system design and operation available to the homeowner.

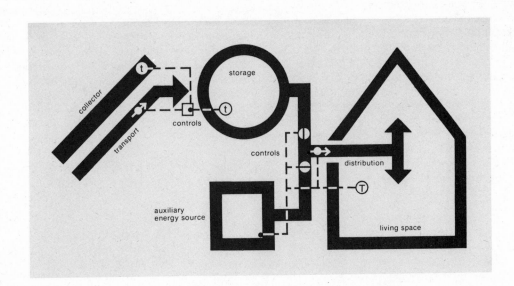

1 Warm-Water Flat-Plate System

Solar heating using water as the heat transfer and storage medium is the most common system in use today. More information is available about the behavior of water systems than about either air or passive systems.

The basic components of a typical water system consist of a collector; storage; a system of piping, pumps, and controls for circulating water from storage through the collector; and a distribution network for transferring stored heat to the dwelling space. The relationship of the various components of a warm-water solar system is seen on the opposite page.

COMPONENT DESCRIPTION AND OPERATION

The liquid-cooled flat-plate collector has a flat absorbing surface integrated with transfer fluid piping which collects both direct and diffuse radiation. Energy is removed from the collector by a liquid flowing through conduits in the absorber plate. The transport fluid is pumped to storage where its heat is transferred to the storage medium (water in this case) and then returned to the collector to absorb more heat. Generally, the transfer fluid is circulated through the collector only when the absorbing surface is hotter than storage (except in instances when snow has covered the collector surface and the heated transport fluid is circulated through the collector to melt it).

Storage consists of either a concrete or a steel tank located near or beneath the building (access should be provided). The tank should be insulated to minimize heat loss. A concrete tank should be lined with a leak-proof material capable of withstanding high storage temperatures for extended periods without deterioration. Heat from the collector is transferred to storage by a heat exchange coil passing through the storage tank. Coil length and size depends on expected collector operating temperatures.

The distribution system consists of a pump and pipes which deliver heated water to the occupied spaces. A thermostat controls the operation of water flow or fan coil unit use in each room or dwelling. Baseboard heaters (convectors) require careful evaluation when not used in conjunction with a fan coil unit. Liquid-cooled flat-plate collectors seldom deliver water above 150°F in winter operation without auxiliary energy or reflected surface focusing. For this reason most warm water distribution systems use fan coil units or enlarged convectors.

Energy is transported away from the collector to storage by water or a water/antifreeze solution. Liquid transport fluids should be carefully evaluated before selection. The liquid must absorb heat readily at various collector temperatures and easily give up heat to the storage medium. Additionally, the liquid should not be corrosive to the system components, toxic, or susceptible to freezing or boiling.

As an auxiliary energy supply, a gas-fired conventional boiler is integrated with the solar system should the solar system fail to function or not meet the dwelling's heating requirement. The distribution piping is run through the boiler where an energy boost may be supplied when temperatures from storage are not sufficient to heat the dwelling adequately.

Domestic hot water piping is run through the central storage tank prior to passing through a conventional water heater. Storage heat is transferred to the hot water piping, thereby either eliminating the need for additional heating or substantially reducing the energy required to raise the water to the needed distribution temperature. The domestic water heating system may operate independent of the space heating system. This is very useful for summer months when space heating is not required.

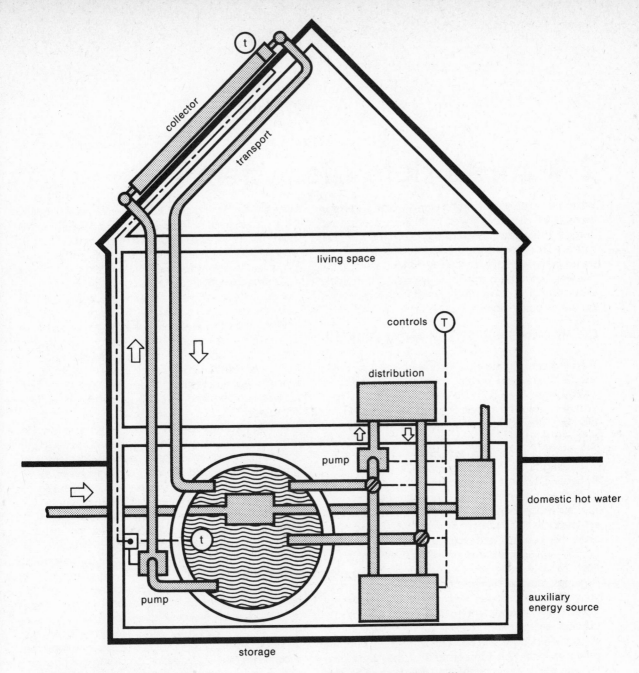

Advantages of Warm-Water Flat-Plate Systems

- They have repeatedly been proved to work well.
- Water is a cheap and efficient heat transfer and storage medium.
- Piping, as opposed to ductwork, uses little floor space, is easily interconnected, and can be routed easily to remote places and around corners.
- The circulation of water uses less energy than the circulation of air with corresponding heat content.
- Much less heat exchanger area is required than with an air system.

Disadvantages of Warm-Water Flat-Plate Systems

- High initial cost, particularly when expensive prefabricated collectors are employed. With the use of large areas of lower-efficiency collectors, the total system cost may be lowered considerably.
- Care must be taken to prevent the occurrence of corrosion, scale, or freeze-up capable of causing damage or blockage.
- Leakage anywhere in the system can cause considerable amount of damage to the system and the dwelling.
- Contamination of the domestic hot water supply is possible if a leak allows treated water storage to enter the domestic water system.

2 Warm-Air Flat-Plate System

Warm-air systems differ from warm-water systems in that air is used to transfer heat from collector to storage. The storage medium can be water, but more typically rock piles are used for warm-air systems. Heat, stored in the rock pile, can easily be distributed to the dwelling space by a forced air system. One possible arrangement of a warm-air system is diagrammed on the opposite page.

COMPONENT DESCRIPTION AND OPERATON

The air-cooled flat plate collector has a solid absorbing surface and collects both direct and diffuse radiation. Energy is removed from the collector by air flowing in ducts beneath the absorber plate. As shown in the diagram, the system may be operated in four different modes:
1. heating storage from collector
2. heating house from collector
3. heating house from storage
4. heating house from auxiliary energy system

The four modes of operation are regulated by several sets of dampers. One set of dampers will direct air flow from the collector into storage or directly into the occupied spaces while another set will regulate air flow from storage to the occupied spaces. The dampers may be adjusted by manual or automatic controls. During modes two and three, an energy boost may be supplied to the warm air by the auxiliary energy system before the air is distributed to the occupied space. The amount of the energy boost is determined by the temperature of the air passing through the auxiliary heater and the amount of heat required at the point of use.

Storage consists of rocks about two inches in diameter, contained in a concrete bin in a basement area or underground beneath the building. The container is insulated on the earth sides to reduce heat loss. The storage capacity should be sufficient to provide several days of winter heating.

Because the temperatures in rock storage are typically highly stratified from inlet to outlet, the air flow providing heat to storage should be from top to bottom. This insures that the temperature of air returning to the collector from the storage is as low as possible, thereby increasing collector efficiency. The air flow, when removing heat from storage, should be in the opposite direction to insure that air returning to the rooms is as warm as possible.

Distribution of the hot air to the rooms comes either directly from the collectors or from storage. The ducting required to conduct the air from the collectors to storage is extensive when compared to analogous piping requirements for liquid-cooled collectors. Two blowers are required to distribute air throughout the entire system.

Auxiliary energy systems of almost any type may be used in conjunction with a solar system. The auxiliary system may be completely separate or fully integrated with the solar heating/cooling system. However, in most cases it makes economic sense to integrate the back-up system with the solar system. This may mean running the distribution component from heat storage to the occupied space through the auxiliary system where an energy boost may be supplied when storage temperatures are low. Heat from storage may also be used in conjunction with heat pumps, absorption units or rankine engines. The heat pump, a device which transfers heat from one temperature level to another by means of an electrically driven compressor, utilizes the solar heat available from storage to supply necessary heat to the occupied space. The advantage of the heat pump/solar system integration is the reduction of electrical energy required by the heat pump because of heat supplied by solar storage. Also, the heat pump is the most efficient device presently available for extracting and transferring electricity into heat.

Domestic hot water piping is run through the rock pile storage bin. The hot water is preheated before passing through a conventional water heater, thus reducing the water heater's energy requirement.

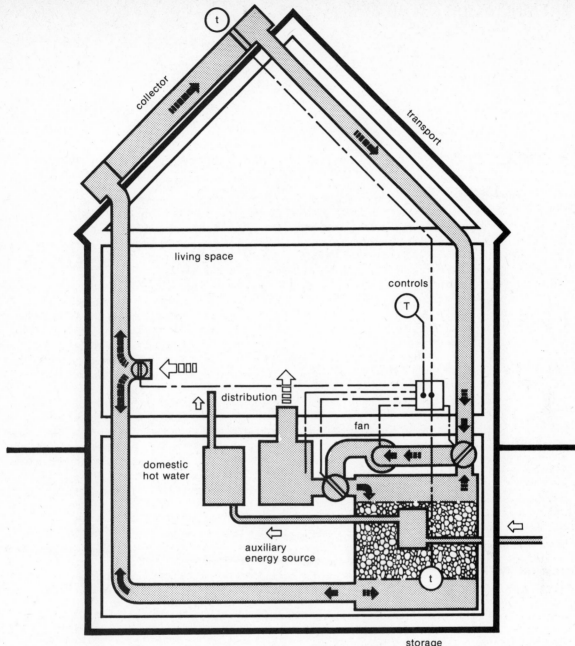

The image contains the following labels: t, collector, transport, living space, controls, T, distribution, fan, domestic hot water, auxiliary energy source, t, storage

Advantages of Warm-Air Flat-Plate Systems

- Capital cost tends to be lower than a water system of the same capacity.
- There is no problem with corrosion, rust, clogging or freezing.
- Air leakage does not have the severe consequences of water leakage.
- Domestic hot water supply is not subject to contamination by leakage from heat storage, as in the water system.

Disadvantages of Warm-Air Flat-Plate Systems

- Ductwork risers occupy usable floor space and must be aligned from floor to floor.
- Air, having a lower thermal storage capacity than water, requires correspondingly more energy to transfer a given amount of heat from collector to storage, and from storage to occupied spaces.
- Air collectors and storage may need frequent cleaning to remove deposits of dust (filters may solve this problem).
- Air systems require a much larger heat exchange surface than liquid systems.

3 Warm-Water Concentrating System

Solar systems with concentrating collectors have not been extensively used for the provision of space heating or cooling. The absence of such equipment from the market, the high cost and uncertain reliability of tracking or concentrating equipment under freezing rain, ice or wind and snow conditions have been the primary reasons for their limited use. However, they do offer advantages over flat-plate collectors — primarily the generation of high temperatures to operate heat driven cooling systems. Representative solar system three utilizes a linear concentrating collector.

COMPONENT DESCRIPTION AND OPERATION

The collector is a linear concentrator with a glass-enclosed pipe absorber. The collector captures only direct radiation and is, therefore, limited to climatic regions with considerable sunshine and direct radiation in winter. However, where applicable the linear concentrating collector offers considerable economies over flat-plate collectors since the necessary absorber area is reduced and the construction of the assembly is often much simpler.

The absorber pipe is a black metal tube within a glass enclosure under vacuum to reduce convection and radiation losses. Radiation is focused on the absorber by a trough-shaped reflector surrounding the absorber pipe.

Storage consists of a steel tank or a lined concrete block enclosure filled with water. Again, the storage unit should be insulated to minimize heat loss. As with most all solar storage techniques, special structural support will be required if the storage tank is to be located in the dwelling.

The distribution system is by heated water to baseboard convectors. Heat is removed from storage by liquid-to-liquid heat exchanger. The heated water is pumped to baseboard convectors located throughout the building. If storage is below a preset minimum temperature the pump continues to operate with a conventional oil or gas-fired furnace assist in the liquid distribution loop.

Transport of collector fluid is by means of a pump which causes the fluid to flow through the absorber, and into the storage heat exchanger from which energy is removed and transferred to storage. The working fluid should be a heat transfer medium which has excellent transport properties and a boiling point above the expected operating temperature of the solar collector.

Domestic hot water piping is run through a heat exchanger in storage, thus preheating the water, before it proceeds to a conventional water heater which also provides additional storage. The water heater may or may not supply an added boost to the water depending on its temperature.

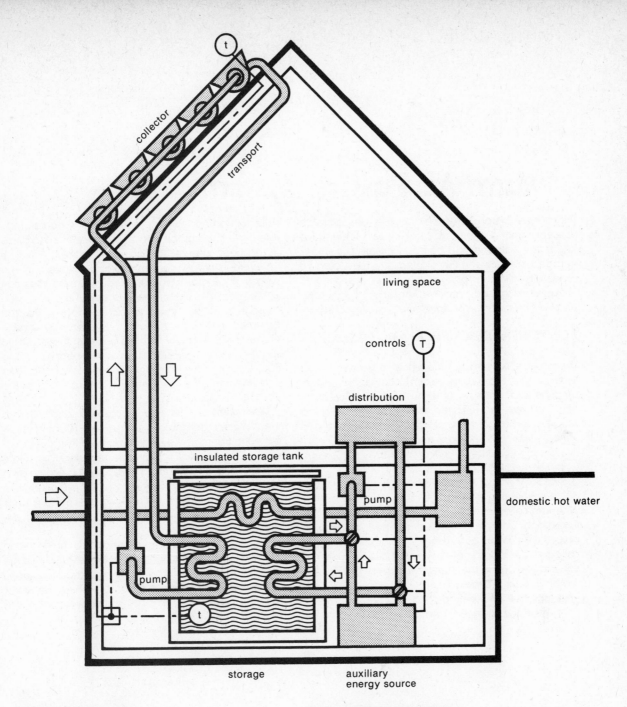

collector

transport

living space

controls ⓣ

distribution

insulated storage tank

pump

domestic hot water

pump

ⓣ

storage

auxiliary
energy source

**Advantages of Warm-Water
Concentrating Systems**
- Potential for more than double the temperatures of either
  air or water flat-plate collectors (particularly useful for
  solar cooling).
- Total absorber area needed is substantially smaller than
  flat-plate collectors.
- Collector forms lend themselves to mass production
  techniques.

**Disadvantages of Warm-Water
Concentrating Systems**
- Capital cost of collectors at present is greater than
  either air or water flat-plate collectors.
- Concentrating collectors may present problems of
  operation, reflecting surface durability, and structural
  mounting.
- Leakage at flexible absorber connections may present
  possible problems.
- Climatic applicability for winter space heating is limited.

4 Warm-Air Passive System

The passive system described here is one possible concept among many. It makes use of extensive south-facing glazing with an intermediate collection/storage wall between the glazing and the occupied space. It relies in part on the thermosyphoning principles discussed earlier. Diagrammatically, the passive system can be represented as illustrated.

COMPONENT DESCRIPTION AND OPERATION

The passive collector, made up of a massive south facing wall of either concrete or masonry separated by an air space from an outer wall of glass, captures direct, diffuse and reflected solar radiation. With the use of automatic or manually operated dampers and vents, the system may operate in four modes:
1. natural ventilation — no collection
2. house heating from collector
3. storage heating from collector
4. house heating from collector and storage
When no collection or heat distribution is required, the vents and dampers may be opened to provide natural ventilation and removal of heat striking the collector. The space may be heated directly from the collector by closing the storage vent duct, thus forcing the heated air into the occupied space. Once sufficient heat has been transferred to the space, the storage vent may be opened and heated air from the collector transmitted to storage. If heat is required at a later time, the storage vent may be opened to allow stored heat to enter the occupied space.

Several storage concepts are employed in the system. The exposed masonry wall which the radiation strikes acts as a storage element. The warmed masonry surface transmits collected heat to the occupied space by radiation. The second storage element is a rock pile located beneath the occupied space. Insulation is placed between the rock pile and floor surface to avoid overheating the space. Water or containerized salts could have also been used as the storage medium.

Distribution of heat to the occupied spaces is from the collector or storage component. Ducting is required to transport the heated air from the collector to storage, and a small fan may be necessary to circulate this air. Heat is distributed to the space by convection from the collector and/or storage, by radiation from the collector and surrounding surfaces, and to a small degree, by conduction from the collector and surrounding surfaces.

Domestic water heating is not directly integrated in the solar space heating system. However, a storage tank or the domestic hot water piping may be placed in the rock pile storage in order to preheat the water before it passes through a conventional electric or gas water heater.

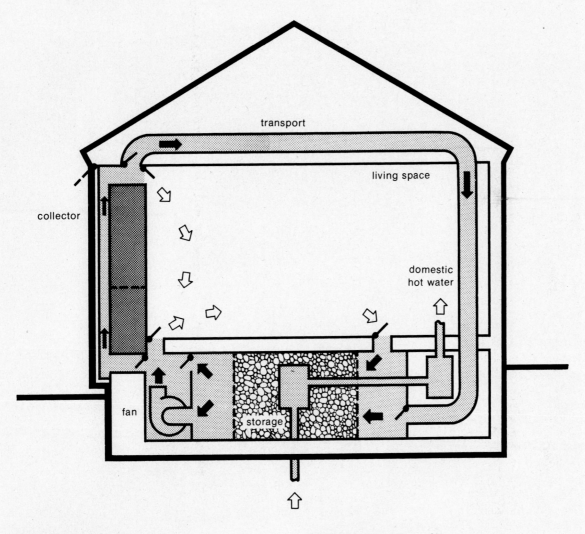

transport

living space

collector

domestic
hot water

fan

storage

Advantages of Warm-Air
Passive Systems
- A system with electrical controls can be designed to operate manually in a power failure.
- Cost should be reduced through simpler technology and elimination of a separate collector.
- Collector serves multiple functions (i.e., can be a wall or roof).

Disadvantages of Warm-Air
Passive Systems
- May not be cost effective relative to warm-air or water flat-plate collector systems.
- In many cases, require automatic or manual insulating devices which are expensive and may require lifestyle modification.
- Larger unobstructed area needed to the south of the house for a vertical passive collector than for a roof collector.
- In some climates and for some passive systems low winter sun angles may be disturbing to the occupants.
- Potential problems of occupant-privacy for passive systems with large expanses of south facing glass.
- Potentially large nighttime thermal losses from collector if not properly insulated.

CALIFORNIA

NEW JERSEY

CHAPTER 3

The Solar Decision

Making the solar decision is no easy chore for any family. In this chapter of the book, we will be reviewing solar projects by individuals and architects who have made their personal solar decision . . . and who are happy they did. One point that should come through very clearly is the individual involvement in each of these projects. We feel it is the involvement aspect that subsequently led to an improved living environment for these solar pioneers. Hopefully, the projects on the ensuing pages, representing a fairly good geographical spread across the United States, will be an inspiration to the reader to take a good hard look at what is possible should he be considering solar energy as an alternative fuel for home comfort. The potential is certainly there for a solar application, but it is important that one learn to sort out the wheat from the chaff in terms of a personal project. As in any new technology, there is as much misinformation as bona fide information and, unfortunately, much of this has been dispersed to the public at large. As one attends conventions and expositions, it becomes evident that many manufacturers are falling by the wayside regularly, but there is always another "great idea" to take up the slack. Gradually, the industry will sort itself out. Standards will evolve . . . measuring sticks that will offer the public solid common denominators on which to base their solar decisions. But it will take time, so the watchword for the present must be caution. When you're dealing with a manufacturer, make sure you have some solid evidence that the product will perform as claimed and that some provision has been made for future maintenance should something go wrong. Decisions based on the future are always difficult, particularly when a dollar investment is involved. However, we feel it is definitely possible to make an intelligent decision if you have the basics well in hand. Frankly, that is what this book is all about . . . just the basics. There has also been a lot of dialogue on the concept of "payback." "If I make an investment in a solar system, how long will it be before my fuel savings will equal my outlay?" Granted that is a legitimate question and one that should be analyzed thoroughly. However, to make this a sole criterion for a solar decision is not wise. As we were selecting the projects to be published in this chapter,

we ran into some interesting facets that caused us to take a second look. We uncovered a great many projects across the country. All indications were that most of these were functionally sound. On the other hand, many of these functional systems were what we could call "less than beautiful." It's too bad, but the perpetrators seemed to work so hard at being first on the block with a workable solar system that they let their anxieties override the basic principles of good solid design aesthetics. We feel that the projects in this chapter have that sense of design aesthetics and, as such, are the type of inspirational material worth publishing. Often times a sense of whimsy will appear to creep into the design facet, but that can be a very positive note in the given situation. In many areas of the country, there are tax rebates centered on solar performance and some utility companies offer financial benefits for solar applications. The utility company in the area where you intend to build can be a tremendous source of information, however, most shy away from making any sort of recommendations. Generally speaking, the local utility company (as well as the local library) is an excellent place to begin your quest for information. Energy consciousness is rapidly becoming a universal thing and proper investigation on your part could save you some dollars in the long run. So keep the following points in mind when you are wrestling with your personal solar decision:

- Ascertain the practicality of a solar system specifically for the geographical location and site on which you hope to build a home.
- Get to know and understand the basics of solar technology in order to minimize the possibility of purchasing inferior products.
- Familiarize yourself with the local codes as they relate to solar energy . . . it could affect the scope of your design.
- Investigate tax savings and loan potential for solar applications in the interest of your pocketbook.
- Don't make a decision based solely on the idea of payback; consider future sales potential as well, in terms of added values.
- Don't hesitate to use your imagination in order to create a more satisfactory living environment for yourself. There really aren't too many rules . . . yet!

THE SOLAR DECISION

Wyoming

Awareness of man's need to harmonize his state of living with nature is the impetus behind the architectural designs of Richard Crowther, AIA, and his Solar Group in Denver, Colorado. A pioneer in the development of contemporary architecture, Crowther has incorporated solar and energy-saving measures into his buildings. The home shown on these pages is located near Cheyenne, Wyoming, where special features were necessary to combat high winter winds and a sloping site. For example, the north side is built into the hill and its low profile roof deflects the prevailing winds; windows on the west are protected by overhangs and sidewalls; even the shape of the structure minimizes wind turbulence. As in most solar homes in the northern hemisphere, the southern exposure is critical. Recessed windows on the south exclude the summer sun, yet receive maximum thermal gains during winter months. To the southeast is a protected airlocked entry. Project designer for the house, Paul Karius, was instrumental in the functioning of the solar system. Solar collectors of the air type are mounted on the south-angled roof and cover 460 square feet. The heat is stored in 15 tons of gravel beneath the house, in a 12-foot bed below the atrium. Foil-backed batt insulation lines walls (5½" R-19 type) and roof (9½" R-30 type), forming a vapor barrier around the building envelope. Basement walls are insulated with 2-inch Styrofoam. All windows are double glazed and set in wooden frames. It is projected that 80% of the annual heating requirements will be provided by solar collection and optimized energy conservation. A large percentage of the domestic hot water will also be supplied. Natural ventilation is used for cooling, and is induced by the collector with its fan. Furthermore, 100% of summer hot water needs are expected to be met with this system: hot air collects at the highest point in the house, is piped to the solar collector for further heating, and is then routed to a heat exchanger for hot water heating before being vented to the outside.

Photographer: Karl Riek

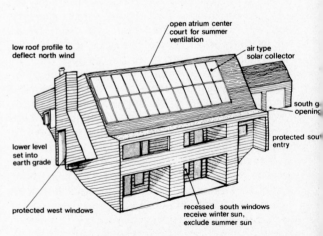

low roof profile to deflect north wind

open atrium center court for summer ventilation

air type solar collector

south g opening

lower level set into earth grade

protected sou entry

protected west windows

recessed south windows receive winter sun, exclude summer sun

Artwork from SUN/EARTH © Richard L. Crowther, AIA 197
Solar collectors supplied by Solaron Corp., Denver, Co.

A sunny disposition belongs to the interior of this three-story home. Pictured above is the west end of the soaring main floor living/dining room. Wide windows look beyond recessed deck area towards the south. Stairs climb to balcony and library, with a bedroom and bath at the other end of the hall. The atrium, pictured above right, provides year 'round greenery and sunshine and may be reached from both the master suite and living room. This large home of 3200 sq. ft. has many fine features above and beyond its solar aspect. Main level extras include study (or playroom, with its hinged-lid window seat), workshop off garage, huge dressing area, and separate laundry room. On the basement level, two good-sized bedrooms face south through recessed windows, as in living room directly above.

First Floor

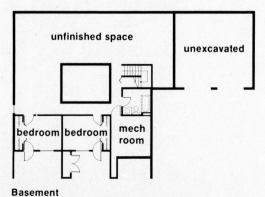

Basement

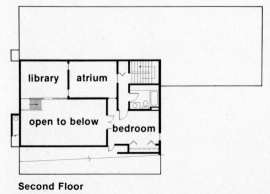

Second Floor

Arkansas

The Delaps, a high school teacher and his wife — on a limited budget — requested that architect James Lambeth design for them a low-cost, high energy-efficient house for year 'round living. What resulted is a technically sophisticated passive design that works with its environment, rather than against it, making comprehensive use of siting, materials, construction and climatic data and integrative design skills. In addition, mechanical simplicity has been attained with reliance upon existing hardware components and operational characteristics. Orientation and insulation were key factors in this design, resulting in a fan-shaped plan with the north wall as the shortest end. A porch extending from this wall deflects cold northerly winds to the sides of the house. The windowless east and west walls are angled out to accept the winter sun and block out difficult summer sun. The major element of the design is a towering wall of windows to the south (below right). These double glazed windows form an 860 sq. ft. passive solar collector. The solar system uses the 12"-thick south thermal wall as a collector and partial storage area. The sun's rays absorbed into the blackened surface of the wall cause heat to be conducted through the wall and then transferred to the interior of the house by convection. The heat storage capacity of the masonry permits continued convection after the sun's rays are gone. The air space between the window wall and masonry wall also heats up by convection. The circulation process can be aided by a thermo-switched fan in the return air duct. The heated air is drawn into return air ducts which lead to a rock bed heat storage area located in the below-ground crawl space. When needed, this heat is distributed throughout the house by a duct system. An auxiliary heat pump provides heat when temperatures of the air in the rock bed storage are inadequate. The interior walls plus a stone fireplace and the underground stone storage provide 488 cubic feet of thermal storage. This supplies enough energy to provide heat through four sunless days.

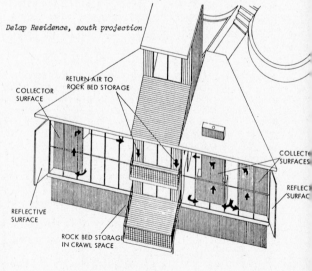

Delap Residence, south projection

COLLECTOR SURFACE

RETURN AIR TO ROCK BED STORAGE

COLLECTOR SURFACES

REFLECTIVE SURFACE

REFLECTIVE SURFACE

ROCK BED STORAGE IN CRAWL SPACE

Photographer: Larry Logan

Situated on an Ozark mountain slope, the 2,000 sq. ft. Delap residence is partly a celebration of their first child's birth, as evidenced by the Mickey Mouse driveway character pictured to the far left — a goodbye to Mickey each morning and hello each evening! The house, which is full of light and space, is oriented to catch the lower winter sunrays and to repel the summer rays with 4' overhangs. The airiness of the interior is accentuated by the contemporary furnishings, as pictured above. North and south openings provide natural ventilation; conventional air conditioning is needed for only two months of the year. The three-bedroom plan is on three levels, with living areas and master bedroom clustered on the first floor; the second and third bedrooms each occupy separate levels.

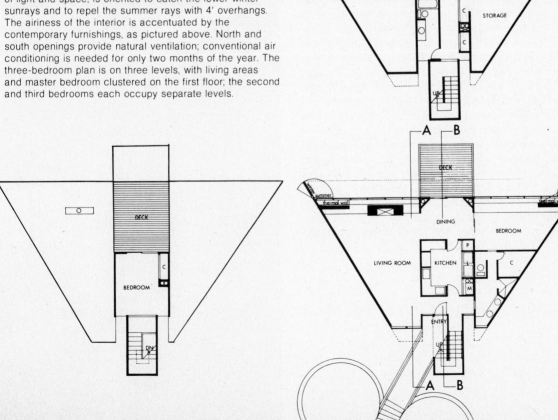

California

This attractive hillside home located on the San Francisco Peninsula gets a plentiful dose of sunshine, ideal for solar collectors. But mild winters, when the temperature seldom drops below 40°F, posed a bit of a problem for a practical backup system. Architect Jim Morelan had designed the house for the site with a westerly roof orientation before the decision was made to "go solar." As a passive energy-saving structure on its own, the house had large amounts of glass on its south and west walls, was carefully insulated, and was partially built into the hillside. Morelan then let the garage, with its rooftop collectors, face south for maximum sun exposure. He and Alten Associates worked out a liquid system, using water in a closed loop design. For locations such as California where water is often in short supply, that "loop" works well — water recirculates through the tank with very little loss from evaporation. The house water goes directly from the main supply line through a heat exchanger in the solar storage tank. The backup problem was solved by installing an air-to-air heat pump that has no connection to the solar system. The heat pump uses only one third as much energy as electric resistance heating. And if there should be an unusually cold spell, there's a second backup in the form of an additional electric resistance coil in the heat pump, activated by a manual switch. So far, though, the owners have never had to use it. Actually, the house has had two sets of owners since its construction in 1975. Quite apart from the solar aspect, the Berrys built according to their lifestyle and philosophy — highly energy-efficient, rooms arranged to encourage family gatherings and good relationships, a strong emphasis on simplicity. In fact, the living-dining room (pictured to the right), not the solar system, was the main feature that sold the second owners, the Judds, on the house. Spacious yet warm, with its stress on wood and earth tones, the room opens to a long deck/balcony. Skylights in the vaulted ceiling brighten the daylight hours and encourage star gazing at night. The modestly equipped kitchen is full of light and charm, with a broad view of rolling hills through an unusual half-circle window. Mrs. Judd loves the pantry: open shelves the full height of two walls provide loads of storage, and a plywood shelf beneath the pantry window doubles as a pleasant desk and menu planning area. The floor plan on page 57 gives an idea of some of the other "extras" in the home. Even before the Berrys decided to sell, people would ask to see the house, somehow expecting it to look different because it was solar heated. For the Judds, the solar aspect was a secondary consideration to the fact that they just liked the house design.

Photographer: Steve Marley

The photo above shows the solar collectors mounted on the south side of the garage. The system uses copper absorber plates with a polyvinyl covering, heat-sealed on the site to an aluminum frame. It yields water at temperatures averaging 120° to 140°F and has a custom-built galvanized steel storage tank with rust-inhibitive paint to withstand temperatures up to 180°F. To the right is the well-laid floor plan of this home. The main level includes a hobby retreat off the living room; downstairs a large storage area is adjacent to the family room. The second story boasts four bedrooms, a utility closet off the balcony, and two efficient back-to-back baths.

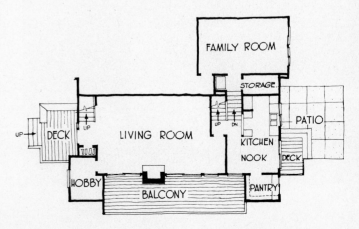

THE SOLAR DECISION

Iowa

Tucked into the woods at treetop level, taking advantage of the view and the sun's warming rays at the same time, is the energy-saving home of Iowa architect, Ray Crites, head of The Ames Design Collaborative. Energy-conscious design has become a mission for Crites and his associates, who compromise neither architectural principles nor energy conservation in their home designs. Sheathed in western red cedar, the core of the house (shown below) is 48 feet long and only 12 feet wide. Acting as passive solar collectors, two greenhouses are cantilevered off the south side of the main structure, adding an extra six feet of width to the living area. Solar heat is gained from this extensive use of windows to the south, allowing the sun to heat the house in the winter. In the summer, leaves of the trees fill out, shading the house. In addition, a simple device, a radiation sensor, is connected to the draperies to automatically close off the glass when it's not advantageous to receive the sun, e.g., during prolonged cloudy times, so as to minimize heat loss. The system is also aided by use of a special fabric developed by the National Aeronautics and Space Administration to line drapes and reflect heat back into the interior when the drapes are closed. The solar heat is supplemented by electric hot water heat. Another important element in the home's design is the lack of windows on the north side of the living area to provide for cross ventilation and to minimize the need for air conditioning. Yet another feature of the system is a special water tank inside the fireplace to help heat water for domestic use and for the hot-water heating system, reducing dependency on electricity to heat the water. Thus the fireplace serves a dual function — as an integral part of the heating system as well as a captivating architectural focal point that visually separates the living and dining rooms (far right). The house also includes zoned heating; there are seven thermostats in the home, so the temperature in rooms can be kept at 50 degrees when not in use.

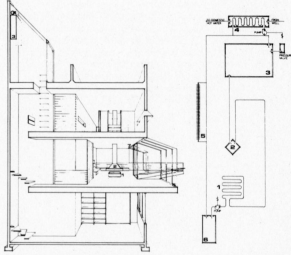

1 Heat transfer pipes in hearth
2 Boiler in firehood
3 Hot water storage tank (winter), solar collector (summer)
4 Heat exchanger for domestic hot water
5 Fin tube for space heating
6 Electric backup boiler
7 Track for double drape system (reflective to interior, absorptive from exterior)

Photographer: Julius Shulman

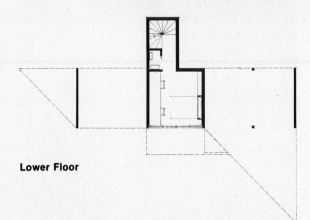

Lower Floor

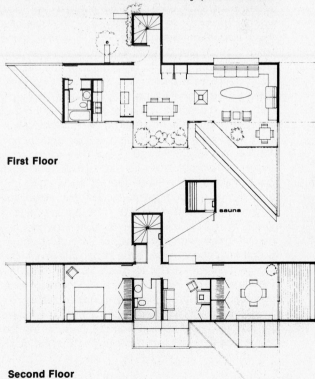

First Floor

Second Floor

A feeling of remoteness and intimacy with the surroundings were requirements set forth by the Crites when formulating designs for their home. In addition, the interior was to be designed around the needs of two professional individuals calling for one bedroom, two studies (convertible to guest rooms), a gallery and television area, two bathrooms and a living/dining/kitchen area. Operational cost was a major concern with electricity, the only readily available fuel. Here can be seen the three-level floor plan that did indeed meet the desirable criteria for the interior, and the picture on the far left shows the wood-framed house perched above a ravine, attesting to its natural involvement with the outdoors.

THE SOLAR DECISION

Ohio

Several years before they even considered solar assistance, E. A. Schmitt and his associates were experimenting with energy efficiency. When the focus shifted to exploration of solar possibilities, they realized that in the Cleveland area a passive system would be preferable from the point of view of dollars and efficiency. Out of this consideration grew the idea of a SUN HOUSE, where they simply built the structure around a solar enclosure — in this case an atrium. The atrium is central to the SUN HOUSE in several ways beyond its obvious location. Since heat loss is extensive through glass — even the insulating type —the house wraps around the atrium with no windows directly to the outside. All glass (mainly sliding doors) orients directly to the atrium which, in turn, has a transparent roof to permit the sun's rays to enter. The concrete floor is painted black for maximum heat absorption. Central to the area itself is a specially designed fireplace: the combustion air is drawn from the outside and takes no warm air from the room. As every room is exposed to the atrium, not only does one experience the visual pleasure of the fire but also its physical warmth. The photo to the right shows just how versatile and attractive an area the atrium is — a year 'round patio, a gardener's delight (temperatures never drop below freezing), light, airy and spacious. The vaulted ceiling lends both drama and elegance. Geometric wood patterns in the furnishings, as well as on the walls and across the atrium roof, add interest and harmonize with the pervading feel of the outdoors. Multi-level (and multi-use) counters, benches and seating are informal and flexible. What fun to have a shirtsleeve cookout when it's zero outside or to toast marshmallows, while three feet of snow cover the ground! And, in addition to its aesthetic value, the atrium supplies from 40 to 60% of the house's heating needs. The graph on the opposite page indicates that the atrium temperature remains considerably warmer than that of the outside, even during the night. The interior glass is exposed to much warmer temperatures than in a conventional house and therefore loses less heat. When the atrium warms above 72°F, patio doors throughout may be opened to heat the house. The concern might be: too warm in the summer. The designers discovered that by opening all of the doors

at the rear of the atrium and all eight windows overhead it was actually cooler than if it were simply an outside patio. It's important to remember that, first and foremost, the house itself is energy-efficient: the exterior wall has full thickness insulation in the cavity; the house is sheathed with 1" Styrofoam, plus 2" Styrofoam at the perimeter of the foundation; the attic contains 10" of blown insulation. Those features, plus the special fireplace design and the solar gain from the atrium construction, are what make heat savings possible. The SUN HOUSE, with its rustic, eye-appealing exterior, versatile and efficient interior and workable passive solar design, is one that may be purchased as a house plan. If you turn to page 93 in Chapter 4, you'll find out more about it.

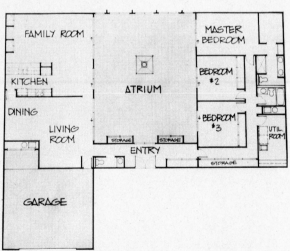

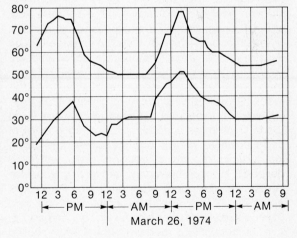

March 26, 1974

Those areas in the floor plan (above) which are shaded —
the garage and the atrium — are not heated. Two electric
heat pumps are located within the house, one in each wing.
The graph (above right) shows inside and outside
temperatures during two days in March. The top line
indicates the atrium temperature — noticeably higher than
that outside, even during the night. These were sunny days
and the effect is evident. For example, at 3 p.m. on March
26th, the outside temperature was 51°F and the atrium
temperature was 78°F.

Tennessee

Architect Lee Porter Butler has taken a somewhat radical approach to passive solar systems in this western Tennessee house. His intent was to design a structure that simply did not get uncomfortably hot or cold; nor did it require a heating or cooling system of any kind. The process involved in bringing such an entity into being was quite complex, demanding an understanding of the total picture — climate, site, building components, and all the elusive aspects that have a bearing. The result was a modestly priced, two-story structure with a greenhouse, or sun room, built onto the length of the south side. The sun room is central to the design. Acting as solar collector and barrier zone, its 360 square feet of glass captures heat, reduces infiltration, and cuts heat loss from the inner glass wall by about 75%. Once the heat penetrates the living space, it is confined from top to bottom as in a well-insulated envelope, with no windows on the north, east or west walls. Another feature to Butler's design is the insulated duct system beneath the structure: heat is stored mainly in the concrete slab, with 18" of dirt beneath that, and a urethane foam layer further below (see cross section on page 63). Air ducts in the dirt permit fresh air circulation and are especially beneficial during the hot Tennessee summers. Unwanted hot air escapes through vents on the roof, while welcome earth-cooled air flows through an underground pipe and is forced up into the living space. Cooling is helped along by an abundance of deciduous trees which shade the greenhouse during the summer months. In addition, direct sunlight is deflected by roof overhangs, preventing excessive penetration. The happy result during the first year of operation: dry, summer temperatures of 75°F and comfortable winter temperatures of 68°F and above. And all that with no mechanical heating or cooling system at all! Even in its bare bones state, the sun room (pictured to the right) has a certain fascination. Its potential stretches as far as imagination permits. As is, its 320 square feet of floor space provides relaxing winter play areas for the harshest of days. For much of the year — recreation room, enclosed "patio," bonus entertaining space, overflow sleep spot for those rained-out nights beneath the stars . . . plus indoor gardening opportunities galore! Butler's concept of heatless houses is not limited to new buildings. Most existing homes can be modified to greatly reduce (and in some cases eliminate) all heating and cooling energy requirements. The cost of such a retrofit can be justified, not only for the sake of energy savings — and saving energy — but also because a greenhouse or sun room is both a property plus and a psychological bonus. In accordance with Mr. Butler's thoughts and designs, we feel that passive solar home construction is a very viable way to combat the depletion of our energy resources.

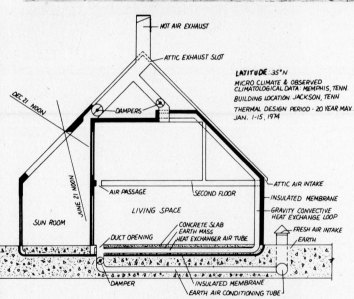

HOT AIR EXHAUST

ATTIC EXHAUST SLOT

LATITUDE: 35°N
MICRO CLIMATE & OBSERVED
CLIMATOLOGICAL DATA: MEMPHIS, TENN.
BUILDING LOCATION: JACKSON, TENN.
THERMAL DESIGN PERIOD - 20 YEAR MAX.
JAN. 1-15, 1974

DEC 21 NOON
JUNE 21 NOON

DAMPERS

AIR PASSAGE   SECOND FLOOR

LIVING SPACE

SUN ROOM

CONCRETE SLAB
EARTH MASS
HEAT EXCHANGER AIR TUBE

DUCT OPENING

ATTIC AIR INTAKE
INSULATED MEMBRANE
GRAVITY CONVECTIVE
HEAT EXCHANGE LOOP
FRESH AIR INTAKE
EARTH

DAMPER
INSULATED MEMBRANE
EARTH AIR CONDITIONING TUBE

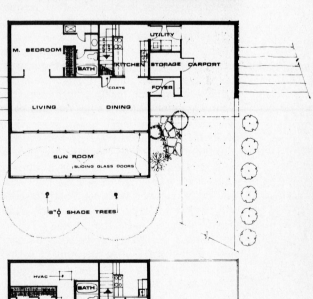

M. BEDROOM   UTILITY

BATH   KITCHEN   STORAGE   CARPORT

COATS   FOYER

LIVING   DINING

SUN ROOM

SLIDING GLASS DOORS

8"ø SHADE TREES

HVAC

BATH

STO.   STAIR   STORAGE

HALL   DOWN

BEDROOM 2   BEDROOM 3   BALCONY   PLAYROOM

UPPER PART OF SUN ROOM

The schematic above shows details of Butler's system,
which he calls an Integrated Natural Energy Environment.
Notice that the angle of the roof provides maximum
exposure to the winter sun, while minimal sunlight enters
at the height of summer. The floor plan to the right gives a
feel for the house itself. The first level has 908 sq. ft.; the
second story, 534 sq. ft. Sliding glass doors divide the sun
room "barrier zone" from the main living areas. Bedrooms
and playroom above look down on the sun room.

New Jersey

Deciding to design a Trombe-type solar heated and cooled home was architect Doug Kelbaugh's first step in a six-month thought process that eventually led to his building and finally occupying the house shown here in Princeton, New Jersey, with his wife and young son. Modifying the concepts of an architectural group in Odeillo, France, Kelbaugh used the south-facing concrete Trombe wall, behind glass, as the constant around which the design process was organized (above right). The basic Odeillo house was modified to include the second floor (with three bedrooms), the greenhouse with basement, fans at the eave, barometric dampers and summer shading from deciduous trees rather than roof overhang. The house was designed to be placed on a 60 x 100-foot lot (with 5 feet of the north property line and 13 feet of the sidewalk) to maximize available sunlight, to save a tree (the largest on the street), and to provide a large single outdoor space. Leaning against and blending architecturally with the Trombe wall is a standard greenhouse. A wide arch leading to the greenhouse is a decorative amenity. On the second floor, the bedrooms are placed along the concrete wall, with separate rooms for toilet, bath/shower and utility along the street wall, which has the only plumbing in the house. Double-height rooms were intentionally avoided to keep costs down as well as to make air circulation easier to predict and balance. Windows were kept to a minimum, especially on the north wall, where they lose more heat than they collect. The north, east and west walls are standard wood-frame with rough-sawn cedar plywood on the exterior and sheetrock on the inside. The cavities — 4½ inches on the first floor, 3½ inches on the second floor and 9½ inches for the roof — were blown full of cellulosic fiber (recycled newspaper). An inch of Styrofoam was placed on the outside of the foundation wall around the perimeter to a depth of two feet. The heat loss of the building by conventional analysis is about 75,000 BTUs per hour. The backup heating system — a conventional hot-air furnace with ducts and registers cast in the concrete wall and one branch that leaves the wall to supply the bathrooms — is separate and distinct from the solar system. Calculations for the first winter in the house showed a 72 percent fuel savings, achieved by letting the indoor temperature swing 3°

to 6° during the 24-hour cycle, thus allowing the concrete wall to collect and discharge its heat. This 2,100-square-foot house cost $55,000 to build, plus the owner's labor —including $8,000 to $10,000 for the solar system. But with the continual escalation of fuel prices, the owner feels it should pay for itself many times over eventually. Some advantages of the Trombe passive solar system include the following: long life, lower operating temperatures, architectural integration into a building, substantial summer cooling, few (if any) fans, pumps and moving parts, ability to work with a variety of back-up systems, and simplicity in building, maintaining, operating and understanding the system.

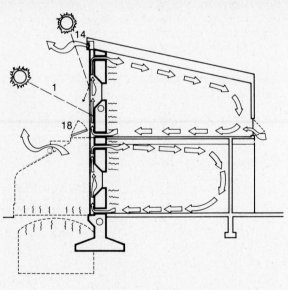

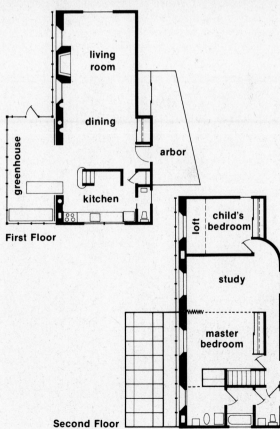

First Floor

greenhouse

living room

dining

arbor

kitchen

Second Floor

loft

child's bedroom

study

master bedroom

The solar heating and cooling diagram above shows the angles at which the winter sun (1) and the summer sun (14) are received by the southern wall (painted a special black to absorb energy) and the flow of heat throughout the entire passive solar system. A solar water heater will probably be placed on the greenhouse deck (18) and adjusted seasonally. A kit which includes 15 colored slides, two blueprints of the solar wall and a six-page description is available for $35 by writing to: Mr. Doug Kelbaugh, 70 Pine Street, Princeton, NJ 08540.

Arkansas

Architect James Lambeth, AIA, commenting on his design of the residence in Fayetteville, Arkansas, shown on these pages: "From my bedroom window, I look down on the McKameys' place. It had to be right. Now, through the trees, the mirrored sky and clouds of the passing day unwind below me. Bill Marlin once said I could shave by it in the morning . . . they'll never know." On this particular home design, the north face takes the brunt of the winter wind and cold. In the summer, the rising and setting sun spends more time on the north facade than the south. The main feature of the house is, in fact, the highly insulated mirrored north wall, which is two and one-half stories high. In its slanted mode, it very effectively deflects the southbound wind. The summer sun is reflected off it also. All of the mirrored panels on the north face are approximately 1½' x 6', with silver mirrored clerestory glass at the highest level, which allows light into the upper zones of the home. The south wall of the home is also all insulated glass, but is clear rather than mirrored. The east and west elevations are left unopened, not only to insure privacy, but to shade the interior from harsh summer sun and sunset. The south side overhangs are wider as a protective element from the sun. The home has a standard backup heating system . . . an electric heat pump sized to function as one normally would in that particular climate. The fireplace is also designed as part of the heat storage system. The home was one of the first James Lambeth designed. The husband of the young couple involved was an electrical engineer, associated with a nearby electrical company and was particularly concerned about the overall efficiency of this passive solar design. From every indication, the design has "worked" and fuel bills have been fairly low. Sitting proudly on its hillside site, the home has cedar board siding, lightened with a bleaching oil which, over the years, has given it a silvered, weathered look. The roof, which cannot be seen, is white composition shingle that effectively reflects the sun.

Photographer: Larry Logan

Photographer: James Lambeth

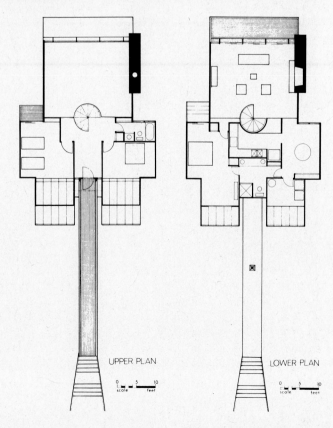

UPPER PLAN

LOWER PLAN

0 5 10
scale feet

0 5 10
scale feet

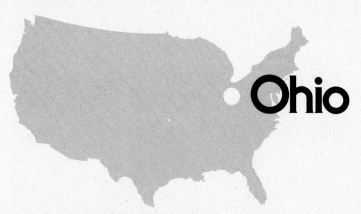

Ohio

Rick and Toni Creveling live in Northeastern Ohio. They didn't set out to build a solar home. However, as Rick researched their homebuilding project, the concept of using solar energy emerged. The Crevelings were aware that a water-to-air heat pump is basically more efficient than air-to-air, so they decided to tie it all together with solar-heated water for a forced-air system for the house. After studying a variety of heat pumps, he decided on a model that cost less than $1000. More expensive units could get heat out of water down to 35°, yet the unit he selected could only handle incoming water down to 50°. The price variation was so high, however, that they rationalized they could put more collectors on the roof to insure that the water stayed at 50° or above. After checking with nearly 100 solar manufacturing firms, they ran into one that was manufacturing on a limited basis, but also selling the absorber plate alone for a very competitive price. The product was just perfect for the system planned, which collects solar energy at low levels. The Crevelings' heat pump operates between 50° and 90°, which they have come to find more than adequate. They realized that more expensive plates would provide water at 150° to 180°, but that seemed more than they needed. Their need for air conditioning is a mere 10 percent of their need for heating, so they were less than concerned about the heat pump's capabilities in terms of air conditioning. It will help, however, and the roof overhangs and insulation help to minimize heat gain in the warmer weather. For storage of the solar heat, they built a 2000-gallon concrete tank into the foundation of the house. Even with the tank's heavy insulation, some heat is lost, and the Crevelings wanted it lost into the house. Most of the tank is below the basement floor level, so if there is any leakage, it will run out beneath the house and down the hill. The tank will store about five or six days' worth of heat to cover sunless periods. If it runs out, their next resort would be their heat-circulating fireplace and system of fans.

Photography: Sterling Roberts and Stev

The Crevelings are extremely energy conscious, and that was a prime consideration when they began planning their new home. Their wall framing, for example, is two-by-sixes on 24-inch centers. That was an economy move in that it utilized less labor and it also nicely accommodates six inches of insulation. There are only 150 square feet of window space in the home, less than 10 percent of the floor area. All are double-pane glass and have wood frames to minimize heat loss. Storm windows assist in winter. They planned placement of their minimum fenestration thoughtfully to insure that the lack of glass did not create a boxed-in look for the interior. The home also has a heat-circulating fireplace with a system of fans. It draws cold air from the outside for burning. Their conventional backup heating system is electrical resistance, which they would use only as a last resort.

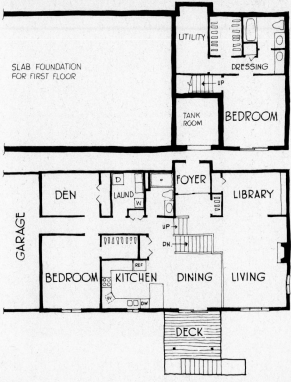

SLAB FOUNDATION
FOR FIRST FLOOR

UTILITY

DRESSING

UP

TANK ROOM

BEDROOM

GARAGE

DEN

LAUND

D

W

FOYER

LIBRARY

UP

DN.

BEDROOM

KITCHEN

REF

DW

DINING

LIVING

DECK

California

Visual excitement is an important dimension to this solar home near Del Mar in southern California. Strong horizontal lines to the natural siding are varied by vertical accents and a definitive diagonal in the forefront. Even the potted plants and brick driveway carry out the sand and earth tones of the surroundings. The site itself is on rugged, Torrey pine-studded terrain that extends halfway up the face of a sandstone cliff, creating difficult problems that had to be overcome in designing the house to accommodate solar heating capability. Architects Gluth and Quigley fit the house into a natural niche of the cliff on the back third of the lot, providing owner, Dr. Andrew Cohen, with a panoramic view of the ocean. Because of the configuration of the land and the close proximity of trees to the house, the garage was designed to accommodate solar collectors. As the photo below demonstrates, the garage sits on a relatively level portion of the lot. The 30° angle of the roofline permits placement of the collectors facing south while main-

taining the aesthetic appeal of compatibility with the overall house. A note of interest: this design recently received an award in a local AIA competition. In the lower right photograph, Dr. Cohen checks the fit of a filon panel that will be used to glaze the solar collectors. A close look shows that wooden runners in the center and along the sides were cut in a scallop pattern to accept the corrugated filon panels. Another layer of panels will later be installed over these in a right angle effect — ridges will lay vertically insuring good water run-off during the rainy season. The roof pitch of 30° from the horizontal is not optimum but adequate for maximum solar exposure in this climate and altitude. Budgetary limitations have prevented total solar installation at this point. The estimate, however, is that the system — 500 square feet of solar collectors and a 1200-gallon storage tank connected to a hydronic heating system — will eventually fulfill about 80% of hot water and space heating needs. Technical advisors were B & A Engineering, Del Mar.

Photography: George Lyons

California

This rustic, simply-styled house nestled in the Redwood forests of Mendocino County uses no fossil fuel for heating either the interior space or the domestic hot water. Built in 1977, the 950 square-foot Stearns house sits in a climate that typically offers foggy, damp mornings and evenings, with sunshine for much of the day. Architect Paul Timberman and physicist/engineer Ralph BruinSlot, of Sun House Design, joined talents and environmental concerns to bring forth their concept of solar energy conversion. Their design is intended to harmonize the outside environment with the building itself, while analyzing the local climate and energy needs of the structure, so as to develop a totally integrated, functionally efficient building. Pictured below is the greenhouse (or Sun Room) entrance that provides the heat gain sufficient to warm the Stearns house without the aid of mechanical heat pumps or fans. Excess heat is drawn off by convection and ducted to a rock storage bin and from there to under the house. The underfloor area is insulated and

creates a hot-air plenum which indirectly heats the living space by radiating through the floor. A vent from the plenum into the Sun Room completes the cycle. A vent by the return vent into the living space allows hot air to be drawn directly into the living space when required, i.e., during early morning. Vents on top of the roof at the ducts remove any excess heat not needed, and vents into the rock storage from outside can introduce cool air into the system at night to provide a natural cooling mode. Since their Spring occupancy, the owners have discovered that the system performs so well that no additional heat (provided by the fireplace) has been required at all. This passive system was chosen because it is less costly than one which requires solar collectors strung along the room. Matching the simplicity of the exterior is the free-flowing interior with the living room opening to the L-shaped kitchen. At the rear of the house are a bath and sauna, plus a loft (not pictured) for sleeping and/or studio space tucked under the roof.

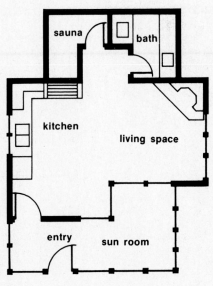

THE SOLAR DECISION

Illinois

Neither Mike nor Ellen Jantzen is an architect or engineer, yet they designed and built their vacation home by themselves. By doing it that way, they were able to construct and furnish the home, including appliances, build a driveway and include a totally controllable solar heating-cooling system for less than $10,000. The solar system is relatively basic, yet it can heat the house to 100° when it's 5° above outside. In winter, the reflective Mylar-lined panels direct solar energy through transparent roof sections into heat-storage tanks under each of the bubble windows on the first floor. Each of the tanks is 16-foot insulated steel, filled with water and anti-freeze. Steel tubes conduct heat through the solution. When the lids of the tanks are closed, they form window seats. The tanks are closed as collectors when the sun goes down and opened later to radiate the heat into the living areas. In the warmer seasons, the roof panels are closed during the day to let the interior heat out. The panels are opened and controlled by a winch. The key to the Jantzens' successful system is insulation. Every surface that divides inside from outside has carefully calculated insulation for perfect temperature control. Outside, the house is painted plywood and corrugated steel, domed with half a silo-top. All interior colors absorb the sun's heat.

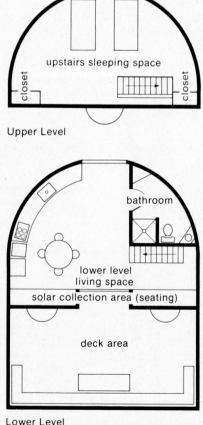

Upper Level

Lower Level

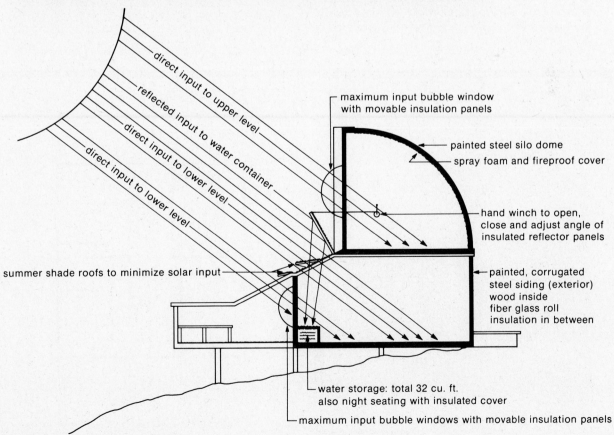

direct input to upper level

reflected input to water container

direct input to lower level

direct input to lower level

maximum input bubble window with movable insulation panels

painted steel silo dome

spray foam and fireproof cover

hand winch to open, close and adjust angle of insulated reflector panels

summer shade roofs to minimize solar input

painted, corrugated steel siding (exterior) wood inside fiber glass roll insulation in between

water storage: total 32 cu. ft. also night seating with insulated cover

maximum input bubble windows with movable insulation panels

THE SOLAR DECISION

New Jersey

This solar design in Old Bridge is the project of Jespa Enterprises. The house features an air-type collector system for space heating and three liquid-type collectors for hot water — together meeting about 60% of heating requirements. A large rock storage bin insulated within the foundation provides for several days' heat carry-over. In summer, the rock storage can be cooled at night to reduce somewhat the temperature of air circulated through the bin the next day. The Jespa house plan itself, prepared by architect Don Watson, is basically energy conserving: 6" stud highly insulated walls; major windows facing south; rooms contained in an efficient rectangle; plenty of natural ventilation. A warm-air return placed at the top of the stair tower recirculates the air from the four-bedroom house and evenly distributes it throughout. A brochure of solar plans designed by Watson and his associates is available for $6.95 from Donald Watson, Box 401, Guilford, CT 06437. Solar equipment was furnished by Sunworks, Inc.

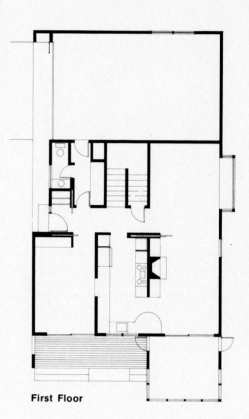

First Floor

Photography: Robert Perron

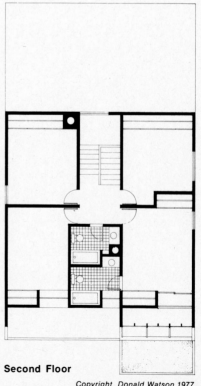

Second Floor

Copyright Donald Watson 1977

74

Connecticut

Solar collectors soak up the sun in this three-bedroom, two-story beach house just a short distance from Long Island Sound. This house was also designed by architect Don Watson, and combines active and passive solar methods for 65% of heating needs. The living room of the 2000-square-foot home is located directly below the collectors, both facing south toward the ocean. As shown below, three banks of roof-top collectors (A) serve the house. The sun heats the water (white arrows) circulated through the collectors. The heated water (black arrows) goes to a 2000-gallon storage tank (B) in the basement, where it circulates through copper tubing. Other tubing connected to the fresh-water supply courses through the tank and is heated to provide domestic hot water. To heat the house, the system circulates hot water from the tank through a fan-coil unit (C). The fan moves air over the coils, heating it, and blows it through ducts to the registers (D). An oil-fired heater (G), and a fuel tank (H), augment the solar system designed by Sunworks, Inc.

Photography: Robert Perron

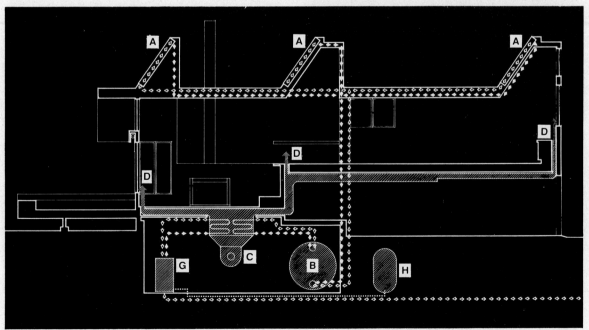

*By permission Donald Watson, **Designing and Building a Solar House***

New York

Architect Harry Wenning, AIA, of Wenning Associates and Solar Structures, Inc., designed this large home on a grant from HUD's Solar Demonstration Program. Total living space is 4000 sq. ft. The home is an efficient two-story box with a finished basement and attic. It is located on a two-acre site near LaGrange, N.Y. and has an excellent southern exposure and protection from the winds by a hill on the northeast side. The entire southern wall on the first floor is heat-collecting glass. On cloudy days, insulated draperies can be drawn to prevent heat loss. The draperies operate on an automatic timer to close at sundown. The active solar aspect is a hot water collection system which converts to forced air to heat the house as well as provide domestic hot water. The 1200 square feet of glass-covered copper collectors are designed as an integral roof system that is both more economical and more attractive. Heat is stored in a 4000-gallon fiber glass tank located strategically under the garage floor. Architect Wenning emphasizes that all construction materials and hardware used in the design are available to any home builder. Two air-source heat pumps are required for backup heating, one for each floor. When solar storage temperatures cannot maintain the desired house temperature, the heat pumps come on automatically to operate the forced-air system. The heat pumps function with outside air temperatures at very low levels, but when they approach zero, electric resistance heaters come into use as a secondary backup system. They also function automatically. While the cost of the dual system for heating and cooling was several times that of a conventional system, operating costs for the home are expected to approximate those for a structure one-quarter the size. The four-bedroom house has cathedral ceilings in the living room, master bedroom, and in one additional bedroom which could function as a den. A heat-circulating stone fireplace with a 5-foot opening is the focal point of the living room, and is functional as well as decorative.

New Hampshire

The Goosebrook Solar Home, shown here, was designed and built by Total Environmental Action, Inc. of Harrisville, N.H. The project was initiated by TEA to examine the cost effectiveness of solar building as well as review the working relationship between solar buildings and local utility companies. This is being accomplished under a National Science Foundation and Energy Research and Development Administration grant. The home features energy conserving design, the use of solar energy for primary heat, and off-peak electricity as a backup system. It is anticipated that the basic fuel bills will be less than one-fifth of homes of standard construction. The high points of the Goosebrook economies are:

• compact design with judicious placement of doors and windows
• extra insulation: 2 x 6 stud walls with R19 insulation, 1" Styrofoam sheathing, 3" of insulation on the exterior surface of the basement walls
• triple glazed windows on east, west and north
• extensive sealing of construction joints, resulting in reduced air leakage
• passive solar heating collection from large

quantities of glass on the south side backed by 6' tubes of water for heat storage
• inexpensive, site-built solar trickle collector system with a combination storage, heat transfer and heat delivery system that increases the trickle collectors' efficiency by being able to utilize low storage temperatures
• a Megatherm off-peak backup heating system, purchased from the Public Service Company of N.H. The Megatherm is actually a large pressurized tank of water which is heated electrically to fairly high temperatures. The heated water serves as stored energy, which automatically supplies lower temperature heat when called upon. Since the tank of water is heated only during the night, during the utility company's off-peak hours (typically between 4 and 8 p.m. in the winter), it is, in effect, less expensive than electricity used during other hours of the day. It effects savings for electrical energy in the short run for the homeowner, and also in the long run in terms of reducing the need for the construction of additional power plants. The plans for this home are offered for purchase in chapter 4, see page 107.

New Jersey

Bob and Nancy Homan built their dream home, a practical solar-heated home nestled in the pines of southern New Jersey. Based upon solar energy plans pioneered by Dr. Harry Thomason, environmental architect Malcolm Wells came up with a design that made a warm environmental statement. Sloped ceilings, massive wooden timbers and exposed beams, and sliding glass doors let in the sun and panorama. Acting as his own general contractor, Bob, a commercial photographer, met frequently with Thomason and Wells before deciding upon the finalized design. What resulted is Solaria, a well-planned, one-floor structure containing some 3,000 square feet — 3 bedrooms, 2½ baths, utility room, kitchen, dining room and living room. It also has a dark room, foyer, two large fireplaces and a photography studio. A loft section on top of the studio is used as an office. A 2,000-gallon-capacity steel water tank sunk below the basic foundation and placed in the center of the structure functions as the solar heat storage depot. When the solar system is on, water from the storage tank is pumped to the top of the 23 solar panels, supplied by Edmund Scientific and installed on the southern side of the house (above right). A large, heavy-duty windmill, supplied also by Edmund Scientific, generates the power to activate the pump. After the water is heated by the sun, it returns to the storage tank, which is surrounded by 30 tons of fist-sized river stones that absorb the heat that the water radiates through the sides of the tank. When air is circulated around the stones, it becomes warm and is forced by blowers through ducts and vents in various rooms of Solaria. The slanted roof on the northern barrier of the house slopes down almost to the ground, allowing cold north winds to blow over instead of through the house; it is also designed to be a garden while adding insulation — two feet of earth coverage to be planted with shrubs and flowers native to the area. Rigid Styrofoam (1½" thick), in addition to the earth covering, offers excellent insulation.

The Homans enjoy cooking together in their dream kitchen (right), rallying point in their solar-heated home. Since they do a lot more cooking than ever before, they appreciate the cushioning in the GAFSTAR sheet vinyl flooring Nancy selected because it resembles actual Delft tile. The modern "country feeling" is carried throughout the house with its mixture of formal and informal touches — living and dining room furniture in cherry woods and velvet chairs sharing spaces with informal beamed ceilings and plants galore. According to Nancy, Solaria was designed for both Bob's and her tastes, which explains the mixture of ideas throughout.

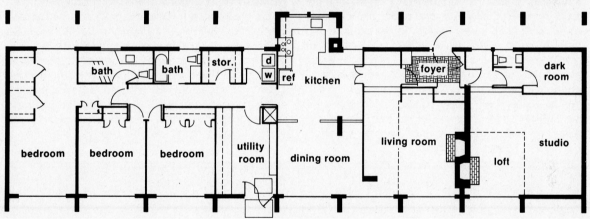

bedroom | bedroom | bedroom | utility room | dining room | living room | loft | studio

bath | bath | stor. | d w | ref | kitchen | foyer | dark room

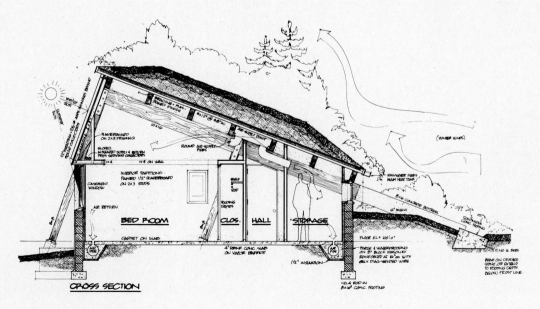

CROSS SECTION

California

The rugged coastal rangelands near Gaviota, California are the setting for this innovative solar home. Jack and Nancy Kittle embarked upon this building venture with Steve Baer of Zomeworks, Inc., Albuquerque, New Mexico. Conservationists in thought and deed, the Kittles liked the multi-clustered "zome" design of Baer's utilizing a passive solar system independent of fossil fuels for backup needs. A free-form shape that is both dome and cube, a zome is a many-sided, vertical-wall building module. The versatile zome house may be strung out in a line or clustered together around an open core. Baer and associate Dick Henry chose a design that used nine modules grouped in a bracelet effect, with twelve south-facing skylights as collectors which are part of a louvered system developed by Baer. Insulated "skylids," installed on an east-west axis, open and close by means of a Freon gas exchange within special cannisters. Sunlight enters the skylights in the morning; when the Freon in the outer can heats, it boils over into the inner cannister, causing an imbalance that opens the louvers; as afternoon passes, the process reverses itself and the louvers gradually close, retaining the heat for the night (and during any cloudy periods). In addition, a manual override allows the Kittles to regulate the amount of heat and light, preventing discomfort during summer months. Woodburning stoves and masonry fireplaces are used for unusual cold spells. The Kittles use electricity only for lighting and appliances, propane gas for the kitchen range. Hot water is provided via flat, roof-mounted solar collectors stored in a triple-insulated water heater. During the building process, prefabricated frameworks for the modules were erected, a plywood skin attached, and between skin and outside sheathing a layer of 15-pound felt inserted. On the inner side of the frames, 4-inch fiberglass batt insulation was covered by gypsum wallboard. Freestanding walls of thick, efficient concrete blocks sandwiched inside the zome walls, and tile floor laid in cement act as storage areas for the heat.

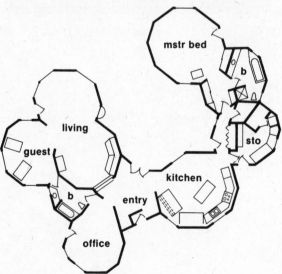

Floor plan to the left shows the nine zomes totaling 2100 sq. ft. looped around a patio that faces south. Living room with built-in sofa (top left) gains spaciousness from double zome construction. Windows throughout are double paned to prevent heat loss. Kitchen cabinets (top right) are unusually shaped to fit the geometric ceiling lines. Below them sits an older range which uses bottled gas; off to the right is a French porcelain stove that burns wood on chilly mornings. Directly above is a close view of the passive "skylid" system. Composition shingles cover the non-solar portions of the roof. Exterior material is redwood from salvaged wine vats, bleached for a weathered effect.

An artist's interpretation of a solar home includes functional utilization of flat-plate collectors for the roof of this home. Libby-Owens-Ford Company's SunPanel solar collector is designed for construction of new homes or for the retrofitting of existing homes.

CHAPTER 4

Solar/Energy-saving Home Plans

Basically, a solar home is simply a weather-tight box that uses solar energy as a source of heat. To do this, the solar house is designed to take maximum advantage of the local environment — orientation to the prevailing winds and sun, the amount of insulation required, and the most efficient use of glass (little or no use on the north side, skylights and greenhouses on the south). In keeping with the times, many architects, designers and plan service firms who specialize in creating pre-designed plans are now offering solar and energy-conserving home designs that you can buy and build. Obviously, no one solar design can be perfect for the wide range of climate and local conditions that can be found across the country. The designers of the solar plans offered in this book have provided efficient, well-insulated homes that can be easily oriented to the proper position for maximum efficiency and have coupled this with what they feel is the best solar heating system for their particular home. In most cases, the designers give the specifications for size and type of solar heating system required, leaving the decisions of the specific brand, the ultimate capacity and the type of back-up heating system required (depending on local conditions), up to you — and a consulting professional (you may need to qualify for a home loan). Some of the plans are designed to use air-to-air systems, some are "passive," some use heated water that flows into a storage tank where it is then converted for heat distribution — the choice is up to you. Whichever you decide is the best for your area, any of these homes can be adapted to suit your location. To gain added flexibility in orienting your home for maximum solar exposure, some designers also offer either mirror or fully reversed plans. Beginning on page 108, several energy-saving home plans are shown. These plans were developed as a direct result of rising utility rates and the ensuing public awareness of the energy crisis. Constructing an energy efficient home can help you minimize the costly economic consequences of this situation. Incorporated into these designs are energy-saving features which enable you to build a home in which the heating and cooling energy needs are greatly reduced (in some cases as much as 60 percent of the needs of most conventional new homes being built today). This has

been primarily achieved by increasing the insulation and reducing infiltration. And, through the use of proven materials and labor-saving construction techniques, an energy efficient home need not cost much more than a conventionally built energy wasteful home. What you get in a pre-designed plans package is a set of working drawings as complete as any you would obtain from an architect for custom services. The plans can be shown to local building officials to secure building permits and zoning variances; they can be submitted to builders for competitive bidding; and, finally, they can be used for actual construction. Some plans include a complete materials list, which details the number and size of lumber, doors, windows, etc. which will be used in construction. Others may include only a quality materials list, which suggests the type or quality of materials to be used, but not the quantity. Choosing the right plan for you and your family won't be easy. You may have a good idea of how your finished home will appear, but when it comes to such factors as good traffic patterns, adequate physical and psychic spaces and construction budgets, the task becomes more challenging. The presentation plans you'll see are simplified versions of the working drawings and are therefore much easier to read. Imagination and visualization are imperative in planning a new home. Your pre-designed plans will be two-dimensional, but you must think three-dimensional. Picture yourself living in the plan under consideration. Select a plan with the needed rooms and square footage, making certain that the circulation pattern or foot traffic doesn't crisscross areas needlessly. Traffic that passes through the kitchen work triangle is just a plain nuisance. Pre-designed plans are commonly accepted by the construction industry nationwide and have been the blueprint to efficient, enjoyable, economical homebuilding for many satisfied American families. They are a valuable means of obtaining good design at a very reasonable cost if you know what you are doing . . . and have the sense to seek advice when you don't, especially in the case of solar homes. The solar industry is growing rapidly and there are many professionals who are well-equipped to help you with any questions or problems you may encounter when building your new solar home.

Small design with solar greenhouse

SOLAR PLAN 6901

- A small design that makes the most of passive solar collection
- "Air lock" foyer for all major entrances
- Wing walls serve as wind shields
- Solar collection panels blend pleasingly with greenhouse windows
- Possibilities for future expansion in garden level basement
- No northern openings help to prevent heat loss
- Adaptable for air-to-air or water storage solar heating systems
- Mirror reverse plans are available if specified

Designer: Continental Home Plans, Inc.

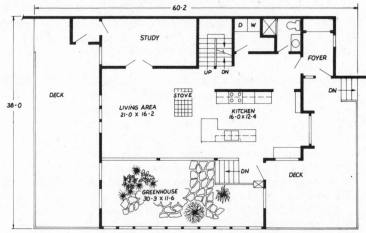

First Floor . . . 1092 sq. ft.   Greenhouse . . . 374 sq. ft.

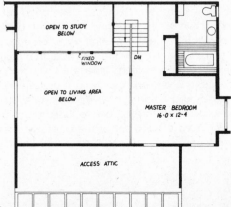

Second Floor . . . 429 sq. ft.

Compact solar contemporary

SOLAR PLAN 6801

- Compact design for a narrow lot
- Water storage solar collection system
- Basement area for storage tank and mechanical equipment
- Master bedroom upstairs has its own private deck
- Sunken living room
- If you live in an area that would require less collector surface, or wish to install the solar panels at a later date, an optional skylight may be installed over the entry to admit more light
- Mirror reverse plans available if specified

Designer: Dave Carmen

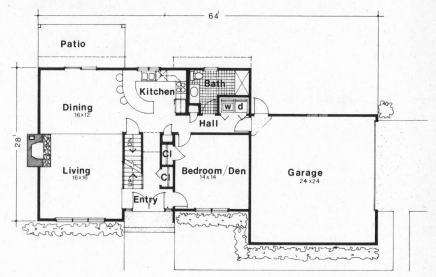

First Floor . . . 1073 sq. ft.

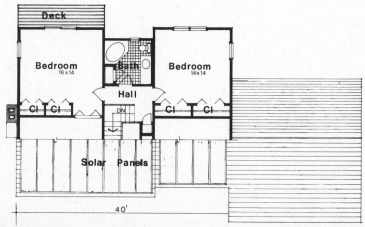

Second Floor . . . 660 sq. ft.

TO ORDER
BUILDING BLUEPRINTS
SEE PAGE 143

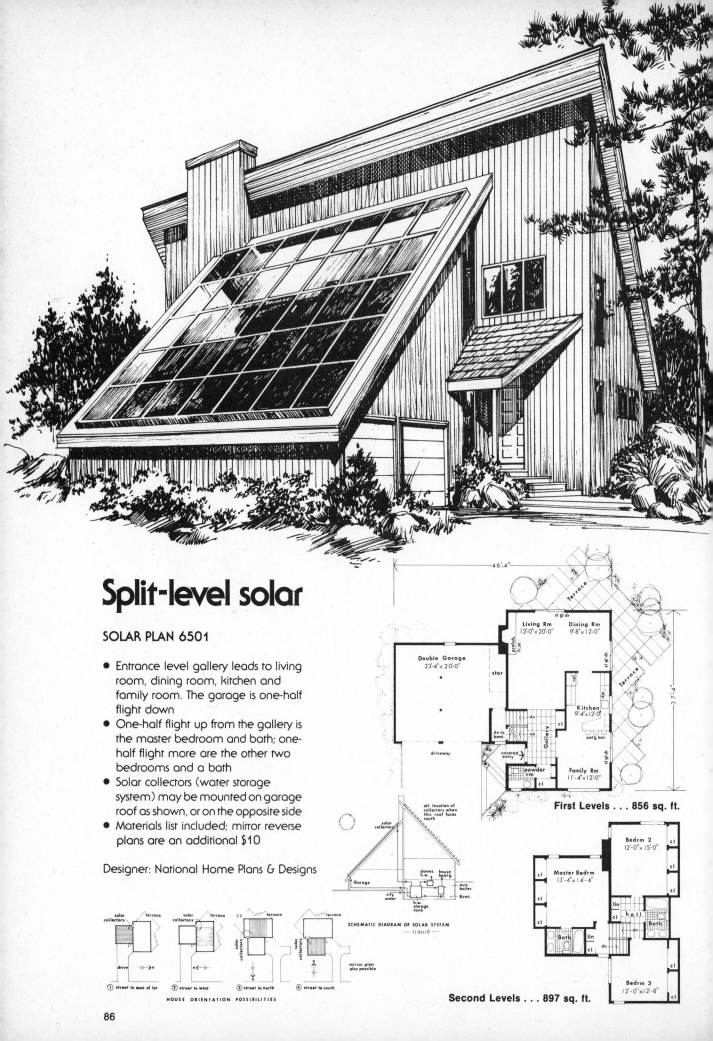

Split-level solar

SOLAR PLAN 6501

- Entrance level gallery leads to living room, dining room, kitchen and family room. The garage is one-half flight down

- One-half flight up from the gallery is the master bedroom and bath; one-half flight more are the other two bedrooms and a bath

- Solar collectors (water storage system) may be mounted on garage roof as shown, or on the opposite side

- Materials list included; mirror reverse plans are an additional $10

Designer: National Home Plans & Designs

First Levels . . . 856 sq. ft.

Living Rm 12'-0"x 20'-0"
Dining Rm 9'-8" x 12'-0"
Double Garage 23'-4" x 20'-0"
Kitchen 9'-4" x 12'-0"
Family Rm 11'-4" x 12'-0"
driveway
covered entry
powder rm
Gallery
eat'g bar
stor
46'-4"
37'-4"
Terrace

SCHEMATIC DIAGRAM OF SOLAR SYSTEM
— liquid —

solar collectors
alt. location of collectors when this roof faces south
Garage
domes. h.w.
house heat'g
aux boiler
cify water
h.w. storage tank
ret.
Bsmt.

HOUSE ORIENTATION POSSIBILITIES

① street to east of lot
② street to west
③ street to north
④ street to south

solar collectors   terrace   drive

mirror plan also possible

Second Levels . . . 897 sq. ft.

Bedrm 2 12'-0"x 15'-0"
Master Bedrm 13'-4" x 14'-4"
Bath
hall
lin
Bath
Bedrm 3 12'-0"x 12'-8"

Walkout ranch has solar panels hidden

SOLAR PLAN 6902

- Spacious design with a built-in office
- Solar panels are located on the rear roof and cannot be seen from the front
- Vaulted ceilings
- Expansion basement for family room, future bedrooms
- Adaptable for air-to-air or water storage solar heating system
- Mirror reverse plans are available if specified

Designer: Continental Home Plans, Inc.

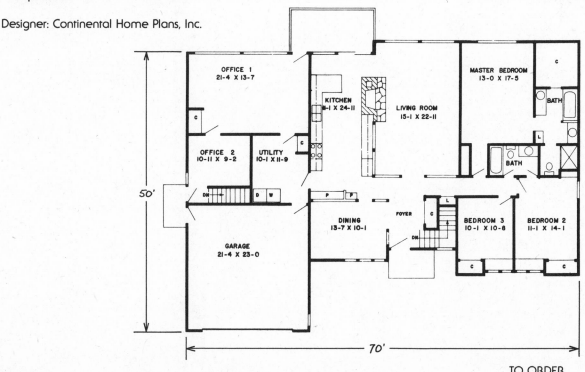

Living Area . . . 2450 sq. ft.

TO ORDER
BUILDING BLUEPRINTS
SEE PAGE 143

Solar salt box

SOLAR PLAN 2341

- Colonial salt box style has collectors on rear roof as shown in rendering
- Four bedrooms clustered on second floor
- Entrance foyer is designed so that either the front or back can become the entrance, allowing as much latitude as possible for proper orientation of the collector panels
- Solar collectors are water storage type with storage tank in basement
- Prefabricated heat recirculating fireplace in living room
- Efficient U-shaped kitchen; laundry room between breakfast nook and garage
- Materials list is included; mirror reverse plans available for an extra $10

Designer: Samuel Paul, AIA

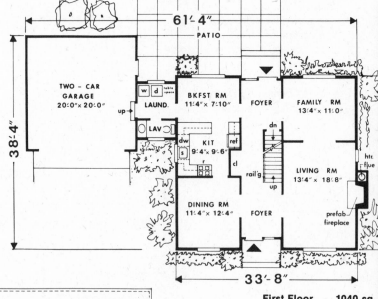

First Floor . . . 1040 sq. ft.

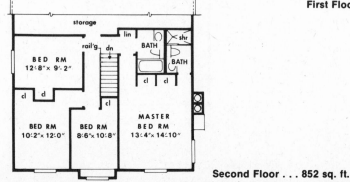

Second Floor . . . 852 sq. ft.

Exciting design with three bedrooms

SOLAR PLAN 6007

- Exciting design with conversation pit, great room, octagonal kitchen and breakfast nook and formal dining room
- Upstairs master bedroom and bath plus second and third bedrooms
- Inner courtyard effectively blocks winds from front entry
- Solar collectors are on ground level (air-to-air system with brick heat storage)
- Solar collectors to left of great room face due south — plan may be reversed for proper positioning
- Mirror reverse plans available if specified

Designer: Michael A. Studer

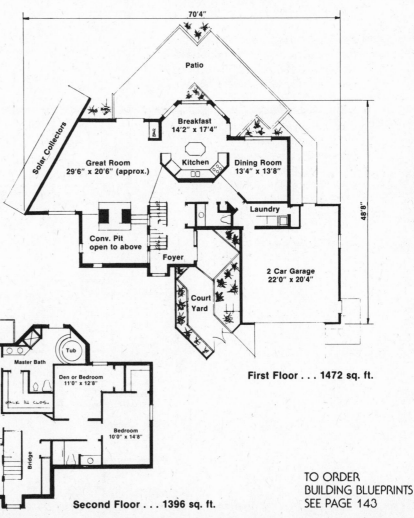

First Floor . . . 1472 sq. ft.

Second Floor . . . 1396 sq. ft.

TO ORDER
BUILDING BLUEPRINTS
SEE PAGE 143

Two-story with room for expansion

SOLAR PLAN 6601

- A compact design with ample room for expansion — finish the upper level and have 1100 sq. ft. of living space, then finish the lower level when you need it
- Mid-level entry helps isolate cold air from the living areas that are a half flight up and down
- Solar heating system is air-to-air
- Plans include mechanical details of solar furnace and ducting system
- Cathedral ceilings throughout
- Materials list additional $10

Designer: Solar Shelter

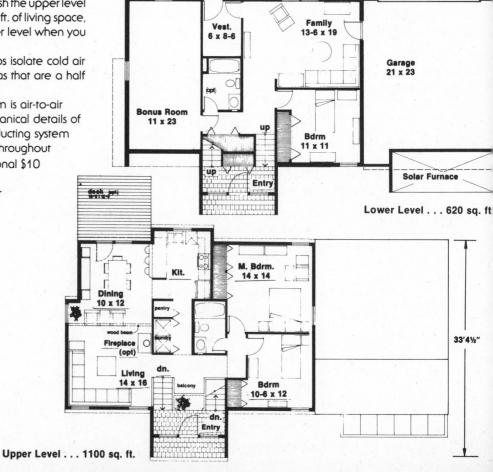

Lower Level . . . 620 sq. ft

Upper Level . . . 1100 sq. ft.

Unique coal stove aids solar system

SOLAR PLAN 6909

- 75% of the heating needs for this home are provided by solar with the balance coming from a unique coal-fired stove located in a special concrete room
- Maximum insulation and absence of windows on the north side assist the heating plan
- Tri-level layout places kitchen and study on the first level, living room on the second and bedrooms on the third
- Generous use of brick highlights the front elevation facing west. Balance of the exterior is wood siding
- Maximum solar collectors number 25 for 487.5 square feet on south roof
- See-thru fireplace, 2 x 6" stud walls, and vaulted ceilings are other features
- Mirror reverse plans available if specified

Designer: Continental Home Plans, Inc.

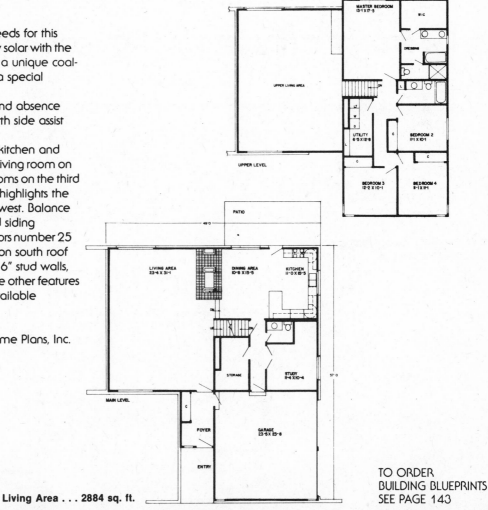

Living Area . . . 2884 sq. ft.

TO ORDER
BUILDING BLUEPRINTS
SEE PAGE 143

Solar is right at home in the mountains

SOLAR PLAN 6905

- Dramatic shed-style roof includes a maximum 28 solar collector panels (546 sq. ft.)
- The maximum number is engineered to provide 85% of needed heat
- Solar rock storage and air handler are located in lower level with hot water preheat tank and electric hot water heater
- Cedar shakes, wood siding and double-glazed wood-frame windows add to rustic appearance
- First floor has large living room with fireplace, circular bar with refrigerator and sink, and opens onto wood entrance deck
- Fully-equipped kitchen includes convenient eating bar
- Mirror reverse plans available if specified

Designer: Continental Home Plans, Inc.

First Floor . . . 1667 sq. ft.

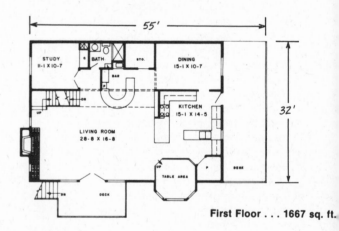

Second Floor . . . 1141 sq. ft.

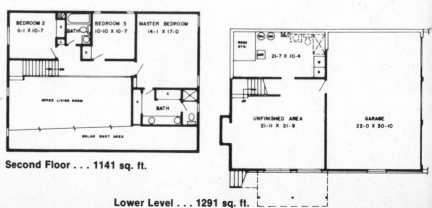

Lower Level . . . 1291 sq. ft.

"Sun House" design relies on passive system

SOLAR PLAN 6701

- Home design revolves around a solar enclosure or atrium which features a transparent roof
- All glass in the home is oriented directly to the atrium, permitting the sun's heat to enter rooms
- Every room is oriented to the atrium which is painted black to absorb heat
- A special fireplace is located in the atrium and draws combustion air from the outside, taking no warm air out of the room
- Entire house is sheathed with T&G 1" Styrofoam
- Full-thick batts used in wall cavities, 10" of blown insulation in attic, and 2" Styrofoam panels cover perimeter of foundation
- Room arrangement places living-dining-kitchen-family rooms on one side of atrium, three bedrooms, two baths on the other
- Mirror reverse plans available if specified

Designer: Edward A. Schmitt, AIA

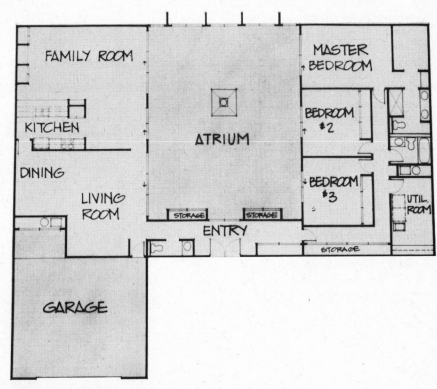

Living Area . . . 2438 sq. ft.
Atrium . . . 1110 sq. ft.

TO ORDER
BUILDING BLUEPRINTS
SEE PAGE 143

Passive design for optimum efficiency

SOLAR PLAN 7001

- Compact split-level design fits on a small lot, contains four bedrooms, two baths, living room, kitchen and dining area
- Solar panels on front roof collect heat, transfer heat to air which flows through wall ducts to basement where the heat is stored by gravel and heats the floor. Shutters and vents control heating and cooling
- A separate collector is mounted on the upper roof to provide hot water
- Two bedrooms are on lowest level; living room, kitchen and dining area and foyer are on second level; two bedrooms and second bath are on third level
- Mirror reverse plans available if specified
- Quality list is included

Designer: Solarama

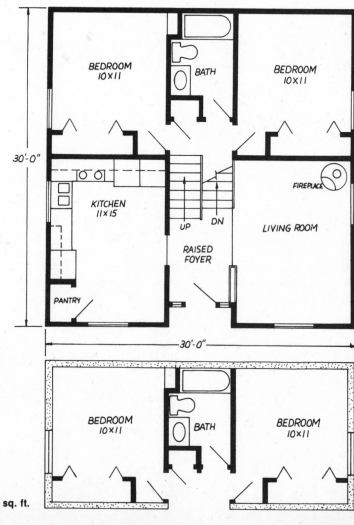

Total Living Area . . . 1320 sq. ft.

Tri-level home emphasizes adult living

SOLAR PLAN 6904

- Shed roof style exterior makes use of southern exposure for roof solar collector panels and highlights fieldstone used on the interior fireplaces
- Behind the two-car garage is solar heat rock storage room, water heaters and air handling system
- An entire floor is devoted to adult living with great emphasis placed on privacy
- Reached via circular staircase, the second (top) floor includes upper living room, study, master bedroom with adjoining deck, and bath featuring a sauna and Jacuzzi spa pool.
- Adjoining the kitchen are a galley-style laundry and lavatory
- Lower level also includes fireplace in family room
- Mirror reverse plans are available if specified

Designer: Continental Home Plans, Inc.

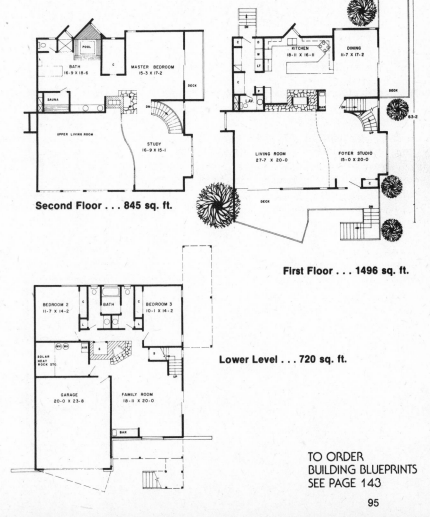

Second Floor . . . 845 sq. ft.

First Floor . . . 1496 sq. ft.

Lower Level . . . 720 sq. ft.

TO ORDER
BUILDING BLUEPRINTS
SEE PAGE 143

Contemporary solar with many extras

SOLAR PLAN 3895
3895-A (with basement)

- Pleasing contemporary design with decks, vaulted ceilings
- Shed roof on all four sides allows maximum freedom on positioning the home on the site and still insures proper solar collector orientation
- Efficient U-shaped kitchen includes handy breakfast bar and pass-through to dining area
- Three large bedrooms and two baths
- Designed for water storage solar heating system
- Plans package includes complete solar information packet with detailed schematics for installing a complete solar heating system
- Full reverse plans are an additional $20; materials list is an extra $20

Designer: Home Building Plan Service

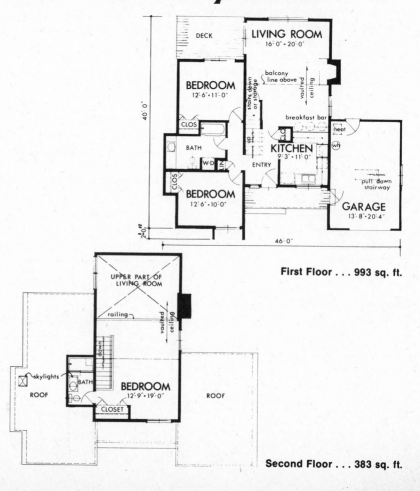

First Floor . . . 993 sq. ft.

Second Floor . . . 383 sq. ft.

TO ORDER
BUILDING BLUEPRINTS
SEE PAGE 143

Great entrance through an atrium foyer

SOLAR PLAN 6903

- Mid-level entry features an open atrium and adjacent closed greenhouse
- Wing walls assist in lowering heat loss from wind effects
- Four large bedrooms plus a study (which could be fifth bedroom), living and family rooms
- Double-entry doors to conserve heat
- Stone hearth and fireplace in living and family rooms
- Adaptable to air-to-air or water storage collection solar system
- Mirror reverse plans available if specified

Designer: Continental Home Plans, Inc.

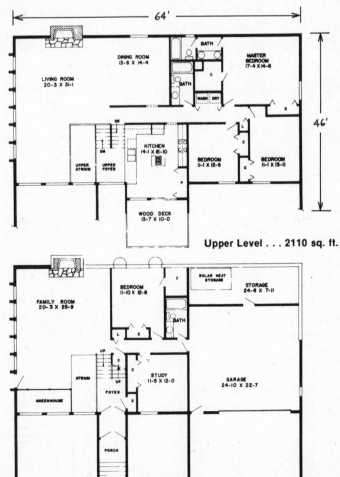

Upper Level . . . 2110 sq. ft.

Lower Level . . . 1300 sq. ft.

Tri-level with bonus space

SOLAR PLAN 6602

- Compact floor plan allows maximum flexibility on orienting home on lot
- Sunken living room enjoys high sloped cathedral ceiling
- Extra large storage area provided
- Third level may be used as bedroom, playroom, den or storage
- Air-to-air solar furnace sits at ground level, uses bricks for heat storage
- Materials list additional $10

Designer: Solar Shelter

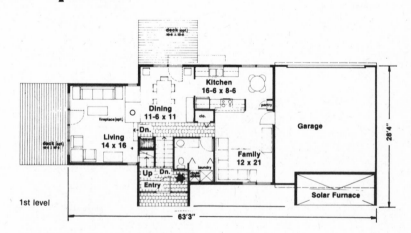

1st level

Total Living Area . . . 1944 sq. ft.

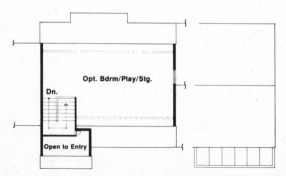

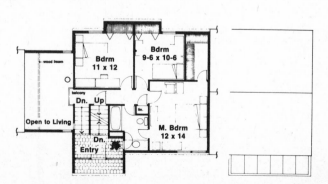

98

Totally passive system aids heating and cooling

SOLAR PLAN 7003

- Self-directed, self-contained system requires low cost to build and maintain. In winter, heat build-up in collector space rises and as cold air rushes in to replace it the hot air goes under the floor and back to collector, heating floor and rocks as it moves along
- Summer heat is vented out roof vents by positioning control shutters, creating a natural venting system by opening the window
- Main level includes three bedrooms, bath, kitchen, foyer, living room and study with built-in desk. Lower level has fourth bedroom, full bath and family room with laundry closet, plus storage
- Fixed glass windows run full width of front elevation, just below roofline
- Mirror reverse plans are available if specified

Designer: Solarama

TO ORDER
BUILDING BLUEPRINTS
SEE PAGE 143

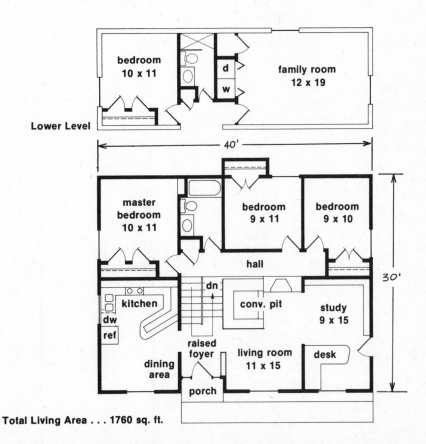

Total Living Area . . . 1760 sq. ft.

Utilizing solar heat on a small city lot

SOLAR PLAN 6907

- Vertical plan allows the collectors to rise above adjacent obstructions commonly surrounding a small urban lot
- Rock storage area for the solar system is located in the basement
- Based on Denver area, solar system provides approximately 80% yearly heating requirement
- Open plan designing locates the living area on the second level with living-dining room, U-shape kitchen, full bath and study
- Lower level has three bedrooms, large bath with space for laundry appliances
- Dramatic slope of the home's roofline continues to the single-car garage with the roof protecting the main entrance to the home
- Mirror reverse plans available if specified

Designer: Continental Home Plans, Inc.

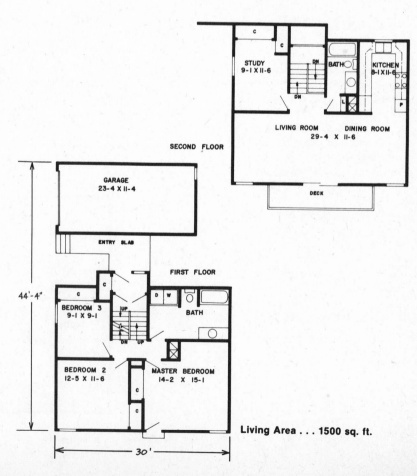

Living Area . . . 1500 sq. ft.

TO ORDER
BUILDING BLUEPRINTS
SEE PAGE 143

Simple appeal in four-bedroom traditional

SOLAR PLAN 6603

- Clean, traditional lines with air-to-air solar furnace at rear of home
- Four large upstairs bedrooms assure privacy and quiet
- Step-down family room with fireplace just off the kitchen and breakfast area
- Large formal dining room behind living room
- Large outside storage area next to solar furnace
- Materials list is an additional $10

Designer: Solar Shelter

Upper level

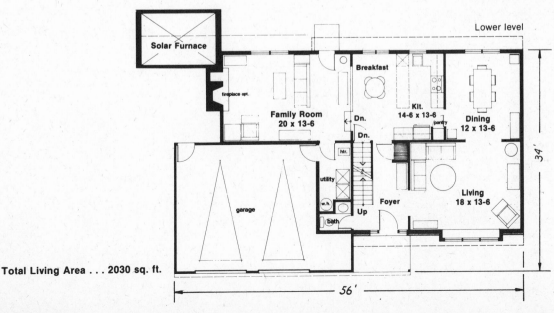

Total Living Area . . . 2030 sq. ft.

Site treatment aids home solar system

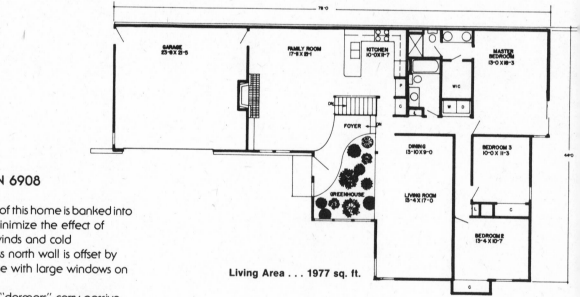

Living Area . . . 1977 sq. ft.

SOLAR PLAN 6908

- North side of this home is banked into a hill to minimize the effect of northern winds and cold
- Windowless north wall is offset by greenhouse with large windows on the south
- Clerestory "dormers" carry passive solar heat to kitchen and family room north areas
- Master bedroom, while on the northeast side, has windows facing south and east, and has optional greenhouse for original construction or later addition
- Total solar system will provide approximately 65% of yearly heat requirements
- Mirror reverse plans available if specified

Designer: Continental Home Plans, Inc.

Contemporary home designed for solar panels

SOLAR PLAN 2148

- Unique plan has highly designed front facade and entry court
- Designed to hide solar panels on the roof
- Three bedrooms in left rear wing away from noisy living areas
- Step-down family room features fireplace and wall of windows to rear
- Built-in bookcases in living room are functional part of exterior design
- Large storage loft in garage adds something extra
- Mirror reverse plans available if specified

Designer: Ron Dick

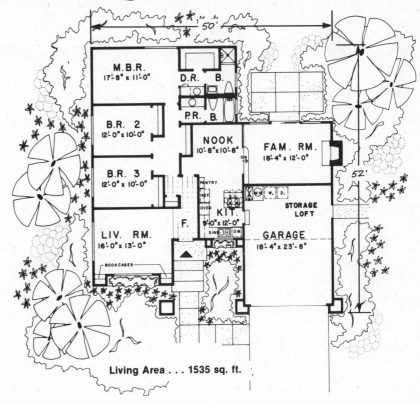

Living Area . . . 1535 sq. ft.

TO ORDER
BUILDING BLUEPRINTS
SEE PAGE 143

Maximizing passive solar collection

SOLAR PLAN 6910

- The configuration of the family room in this home maximizes passive solar collection while providing a beam effect over the entrance atrium
- Sixteen solar collector panels provide maximum 312 square foot area
- Popular ranch styling with a basement
- Kitchen and family room have a vaulted ceiling, plus fireplace
- Master bedroom features compartmentalized bath, dressing, storage
- Exterior combines brick, cedar shakes with vertical wood siding, shake roof
- Mirror reverse plans available if specified

Designer: Continental Home Plans, Inc.

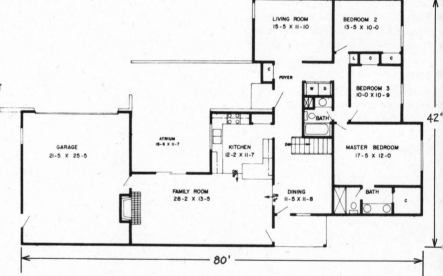

Living Area . . . 1786 sq. ft.

Clerestory windows help heat and cool

SOLAR PLAN 7002

- Split-level "passive" solar design
- Living room connects to mid-level entry; lower level contains the family room, kitchen, dining and breakfast rooms, bath and garage; upstairs level contains three bedrooms and a bath
- Floors are heated by air circulating between the roof collectors and underground gravel surrounding the heat duct
- A separate collector panel over the garage supplies heat for hot water needs
- Mirror reverse plans available if specified
- Quality list is included

Designer: Solarama

Total Living Area . . . 1700 sq. ft.

TO ORDER
BUILDING BLUEPRINTS
SEE PAGE 143

Greenhouse helps solar home system

SOLAR PLAN 6906

- A black stone wall and inexpensive blower system in the greenhouse help to transfer heat into the living areas, assisting the main solar system on the roof. Entry to the greenhouse is off the bedroom wing on the lower level
- Rock storage for the solar system is located off the utility room for close proximity to water heaters
- Placement of the living area on the upper level takes advantage of the views
- Open planning gives rooms added dimension and simplifies traffic patterns
- House has all cedar exterior and is lowered 3' into the ground to assist the solar heating system
- Mirror reverse plans available if specified

Designer: Continental Home Plans, Inc.

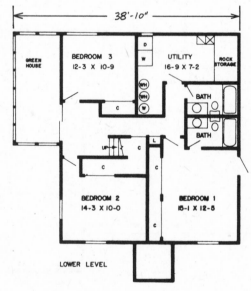

Living Area . . . 2082 sq. ft.

TO ORDER
BUILDING BLUEPRINTS
SEE PAGE 143

Compact solar home design

SOLAR PLAN 7301

- Compact design has judicious placement of windows and doors
- Extra-heavy insulation, triple-glazing and careful sealing of joints, uses normal wood-framed construction methods
- Passive solar heating through large amounts of south-facing glass
- 6' tubes of water store heat which is delivered to house by warm-air delivery system
- Roof-mounted solar collectors
- Three bedrooms, one and a half baths
- Attached greenhouse
- Fully insulated basement
- Mirror reverse plans available if specified
- Materials list included

Designers: Bruce Anderson and Charles Michael

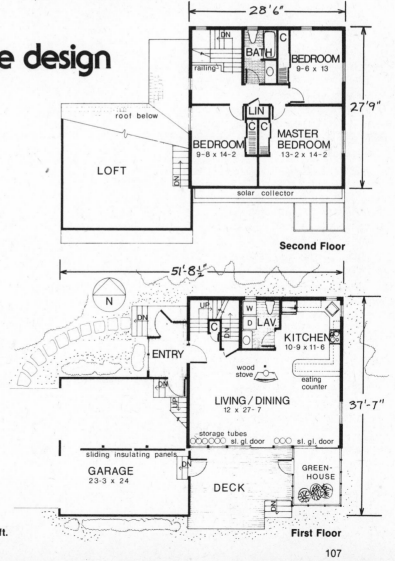

Second Floor

First Floor

Total Living Area . . . 1400 sq. ft.

Rustic country home can be expanded

ENERGY-SAVING PLAN 3334

- A project of the Atlanta Home Builders Assn., this energy-saving home is also expandable
- Interior design features spacious Great Room centered around large woodburning fireplace
- Country kitchen is fully equipped and opens into a formal dining area
- First level also includes two large bedrooms, and master bedroom suite
- Exterior is all cedar with an open air deck and old-fashioned front porch
- Materials list is an additional $10; mirror reverse plans available if specified at no additional cost
- Foundation options: slab, crawlspace or basement. Please specify

Designer: W. L. Corley

Second Floor . . . 730 sq. ft.

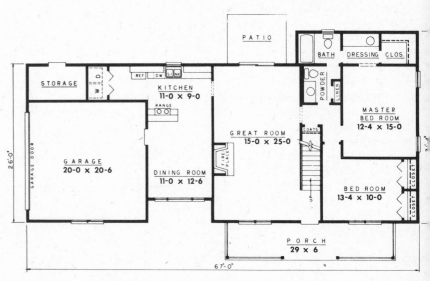

First Floor . . . 1320 sq. ft.

A ranch home with energy-saving features

ENERGY-SAVING PLAN 7103

Full insulation, electric heat pump conditioning, water-saving plumbing fixtures and a vapor barrier system are some of energy-saving features
- Sunken master bedroom has comfortable sitting area above
Other two bedrooms are separated by a full bath
Work center of home includes an efficient U-shaped kitchen with nearby sewing center and separate utility room
Two-car garage has storage space above, workbench area and storage room at floor level
- Mirror reverse plans available if specified; quality list included

Designers: Edsel E. Breland and John L. Farmer

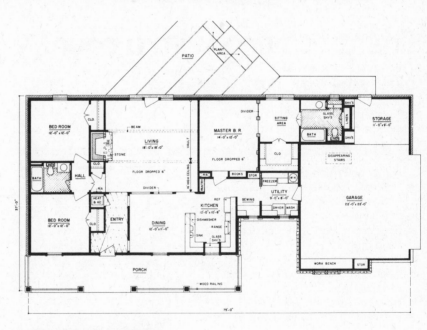

Living Area . . . 1600 sq. ft.

TO ORDER
BUILDING BLUEPRINTS
SEE PAGE 143

Energy-saving with a country flair

ENERGY-SAVING PLAN 7102

- Energy-saving features in this home include 6" wall insulation, 12" attic insulation, exterior vent to fireplace for combustion air, and positive vapor barrier system on ceilings and walls
- Side-entry double garage has space above for future game room
- Country design dining-kitchen areas have a beamed ceiling as does living room
- Heating unit, washer and dryer located in bedroom wing hallway to conserve space
- Mirror reverse plans available if specified; quality list included

Designers: Edsel E. Breland and John L. Farmer

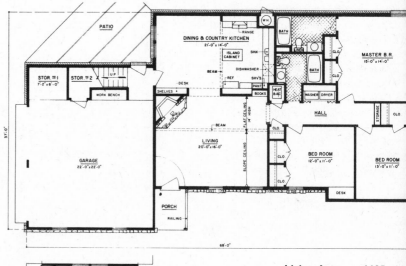

Living Area . . . 1405 sq.

TO ORDER
BUILDING BLUEPRINTS
SEE PAGE 143

Modern two-story provides room for expansion

ENERGY-SAVING PLAN 2260

- Unfinished area on lower level large enough for family room, full bath and extra bedroom
  Main level has three bedrooms above two-car garage-storage area
  Stairways to the upper and lower levels face the main entry
- Efficient L-shaped kitchen includes dining area
- Living room overlooks rear patio-yard
- Full reverse plans are available for an extra $30; materials list is included

Designer: National Plan Service

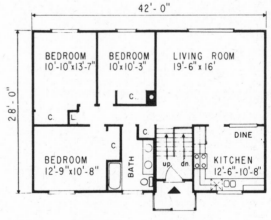

Main Level . . . 1164 sq. ft.

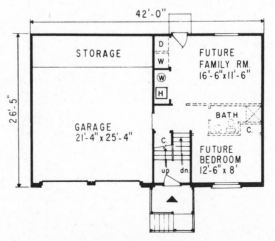

Lower Level . . . 534 sq. ft.

Styled and zoned for family living

ENERGY-SAVING PLAN 7101

- Children's bedroom wing has full bath, more than usual storage and is set apart from master bedroom suite
- Large tiled entry directs traffic to particular areas including dramatic cathedral ceilinged living room
- Efficient kitchen provides immediate access to separate dining room and a family eating space overlooking the porch and patio
- Well-planned utility room adjoins master bedroom, yet is just steps away from kitchen
- Attached two-car garage adds impressiveness to home's exterior
- Mirror reverse plans available if specified; quality list included

Designers: Edsel E. Breland and John L. Farmer

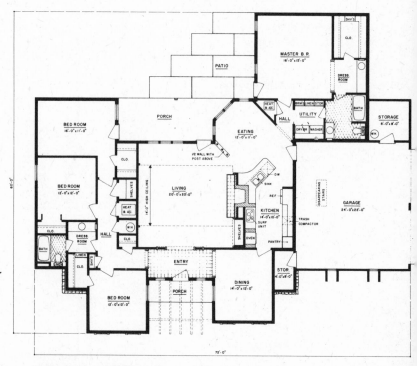

Living Area . . . 2200 sq. ft.

TO ORDER
BUILDING BLUEPRINTS
SEE PAGE 143

Expandable bungalow for a young budget

ENERGY-SAVING PLAN 2240
2240-A (without basement)

- Home can be easily expanded to include additional bedroom, family room with fireplace and wood deck at rear
  Living room opens to an enclosed garden-yard at the side of home
  L-shaped kitchen features corner sink and space for dining
- Bi-fold doors close off a 9 x 10' study which adjoins living room
- Plan is available with or without basement. Please specify
- Full reverse plans are available for an additional $30; materials list is included

Designer: National Plan Service

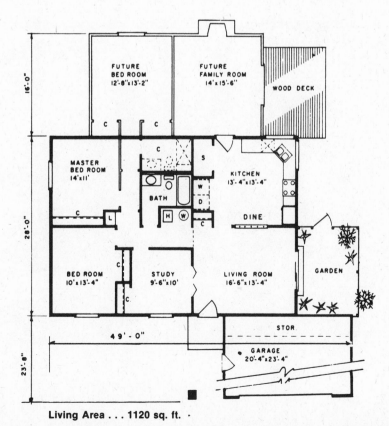

FUTURE BED ROOM 12'-8"x13'-2"

FUTURE FAMILY ROOM 14'x15'-6"

WOOD DECK

MASTER BED ROOM 14'x11'

KITCHEN 13'-4"x13'-4"

BATH

DINE

BED ROOM 10'x13'-4"

STUDY 9'-6"x10'

LIVING ROOM 16'-6"x13'-4"

GARDEN

STOR

GARAGE 20'-4"x23'-4"

16'-0"

28'-0"

23'-6"

49'-0"

Living Area . . . 1120 sq. ft.

Champion Home Builders Co.

Solar Energy Research Corp.

CHAPTER 5

The Solar Shopper

This chapter of the book is, in effect, a mini-catalog of solar and solar-related products currently available on the market. The material is as up-to-date as possible. Changes, however, will undoubtedly have taken place. The product descriptions are based on sales literature supplied to us by the manufacturers. For additional solar product manufacturers, see the index.

For those solar shoppers who choose not to consult professionals for help, we have been given permission to reprint the following guidelines developed for "California Sunshine, A Consumer Guide to Solar Energy," written for the State of California by Interactive Resources, Inc., of Point Richmond, California.

1. Deal only with licensed contractors and insist on a written contract. Any questions or complaints regarding contractors should be referred to your State Contractors' License Board.

2. Ask for proof that the product will perform as advertised. The proof could come from an independent laboratory or a university. You should have the report itself, not what the manufacturer claims the report states. Have an engineering consultant go over the report if you don't understand it.

3. If there is a warranty, examine it carefully. Remember that, according to the law, the manufacturer must state that the warranty is full or limited. If it is limited, know what the limitations are. How long does the warranty last? Are parts, service and labor covered? Who will provide the service? Does the equipment have to be sent back to the manufacturer for repairs? Ask the seller what financial arrangements, such as an escrow account, have been established that will back the warranty.

4. Solar components are like stereo components — some work well together, others don't. If the system you are purchasing is not sold as a single package by one manufacturer, then you should obtain assurance that the seller has had the professional experience for choosing properly.

5. Ask the person who owns one. Ask the seller for a list of previous purchasers and their addresses, and then ask the owners about their experiences.

6. Be careful of sellers who use P. O. Box numbers. Though many legitimate businesses use these outlets as a convenient way to receive bills and orders, a common tactic of the fly-by-night artist is to use a Post Office Box number, operate a territory until the law starts closing in, then move and take a new name in a new territory. Find out from the seller where his place of business is, how long he has been there, and ask for his financial references.

7. Be sure you know specifically who will service the solar system if something goes wrong. Don't settle for a response that the plumber or handyman will do.

8. Don't try a do-it-yourself kit unless you really have a very solid background as a handyman. One or two mistakes could make a system inoperable and you will have no one to blame but yourself.

9. Don't change your use habits simply because you are getting plenty of free energy. Conservation of energy still counts if you want to bring your monthly bills down. Don't blame the seller of a solar heating system if you keep your doors open during the middle of winter.

10. Don't forget your local consumer office or your Better Business Bureau. Both may be able to help you determine whether a seller is reputable or not. Check, too, to see whether there is a local volunteer citizens solar organization who can advise you.

11. If the seller makes verbal claims that are not reflected in the literature handed out, ask him to write these claims down, and to sign his name to it. Compare what he said with what he wrote. Save the statement.

12. If you have what appears to be a legitimate complaint, notify the local district attorney's office immediately, the Better Business Bureau, the local consumer protection agency, or the Contractors' State License Board. Be as specific in your complaint as possible, and give as much documentation as you can.

Collectors and Systems

These Sol-R panels feature a non-metallic frame structure — because metal enclosures may cause heat loss by conduction, the panel frame is of treated and coated wood construction for maximum efficiency with lowest possible heat loss. Panel sizes are 2 x 4', 4 x 4', 4 x 8' and 4 x 12'. Glazing is FRP plastic or 3/16" Crystal Glass. Manufacturer: E & K Service Company, Bothell, Washington.

The del Sol Control, in combination with the March pump, will circulate water or other acceptable liquids in a solar or other temperature dependent system. Two temperature sensitive "Thermistors" are used to determine temperature difference between a solar collector and a storage tank. Whenever the collector is slightly warmer than the storage tank, the del Sol control turns on the pump and circulation begins. As long as the collector can supply additional heat to the storage tank, circulation will continue. However, if the collector temperature should drop, circulation will stop. Circulation always stops, for example, with sudden heavy cloud cover, precipitation, or late afternoon sun. Manufacturer: del Sol Control Corp., Juno, Florida.

Solcan tracking, concentrating fluid collector units are vertically mounted at 45° and face south to automatically track the sun's position from sunrise to sunset. The collectors may be installed on the roof of a structure or on the ground to achieve the maximum angle advantage to the sun. An electronic brain and electric eye control the system's sun tracking capability. The system can work in conjunction with standard forced air or hot water heating systems. Manufacturer: Albuquerque Western Solar Industries, Albuquerque, New Mexico.

This "Hotter Water Faster" collector panel features a new kind of copper absorber plate which increases water heating capacity. And this means greater BTU/hour efficiency for not only more hot water, but also hotter water. The collector panel will heat water or antifreeze solutions for use in domestic water heating, space heating or cooling and process water heating. A universal mounting bracket is available to make installation on any roof possible in minutes. The frame is aluminum; the absorber is copper Rollbond. Adjustable mounting brackets and electronically controlled circulating pump systems are also available from the firm. Manufacturer: Solar Innovations, Lakeland, Florida.

The Sunpump provides a simple and efficient means of solar thermal energy collection for commercial, residential and agricultural applications. The thermal transport and storage characteristics are optimum for heating water which makes the system ideal for heating domestic water and for buildings using hot water heat. Interface with hot air furnaces can be accomplished by use of a secondary heat exchanger. The collection and storage temperatures ranging up to 100°C (212°F) are also compatible with absorption-desorption type air conditioning systems. The unique feature of the patented system is that energy is transported from the collectors in the form of steam rather than the liquid or air used in conventional systems. The focusing collectors are designed so no tracking devices or seasonal adjustments are required. Manufacturer: Entropy Limited, Boulder, Colorado.

The Solaron solar collector for domestic hot water heating is an advanced type of air heating, flat plate collector. The internal manifolding allows the Solaron collector to be completely modular. Factory pre-assembled collector panels are plugged into each other with a minimum of installation time. Air inlets and outlets are field cut into each collector array as required. The collector is designed for installation on any structurally sound surface, such as a roof, wall or specially made supports. Manufacturer: Solaron Corporation, Denver, Colorado.

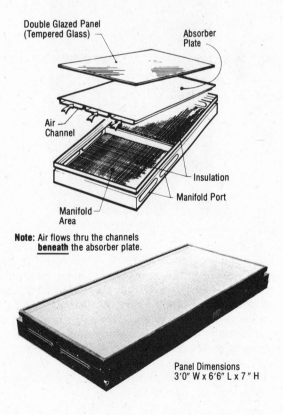

Double Glazed Panel (Tempered Glass)

Absorber Plate

Air Channel

Insulation

Manifold Port

Manifold Area

Note: Air flows thru the channels **beneath** the absorber plate.

Panel Dimensions
3'0" W x 6'6" L x 7" H

Collectors and Systems

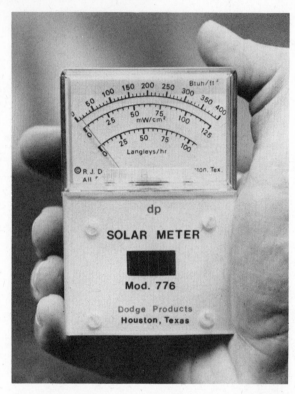

This cut-away view of Libbey-Owens-Ford's new SunPanel flat plate solar heat collector shows the product's two tempered glass cover plates, all-copper absorber plate, three inches of treated fiberglass insulation and all-copper fluid-carrying tubing system. Permanently attached mounting supports simplify handling of the collector during shipping and installation, then function as brackets for attaching the collector to its framing. Manufacturer: Libbey-Owens-Ford Company, Toledo, Ohio.

The Model 776 Solar Meter is a hand-held solar intensity meter used to measure direct plus diffuse solar radiation. The meter is useful to measure the transmission loss through transparent material, and it may also be used to estimate the performance of concentrator systems, lenses, mirrors, etc. To estimate the total insolation over a period of time, several readings can be taken and summed to obtain the total. The total daily insolation is usually expressed in Langleys. The diffuse radiation can also be estimated by shadowing the sensor cell. Manufacturer: Dodge Products, Houston, Texas.

AMSOL's flat plate tracking collector is larger than the average flat plate collector available today (52.7 sq. ft.) and yet is comparatively lightweight since it's made of fiberglass. The tracking device is fully automatic. Manufacturer: American Solar Systems, Arroyo Grande, California.

The proprietary WARM solar system by Solar Energy Research can be utilized in two ways: to provide space heating alone, or to provide space heating and cooling through addition of a solar assisted heat pump. A small increase in cost boosts heating capacity while eliminating the need for a standard heating system to act as back-up. Equipment automatically reverses to cooling mode to maintain comfortable temperatures throughout the year. The system illustrated at left is for a retrofitted Pennsylvania home. Manufacturer: Solar Energy Research Corporation, Longmont, Colorado.

Solarshingles are made of two sheets of glass (three sheets for heating air) separated by glass spacers. The bottom sheet is coated with a black absorber, and water is gravity fed through the shingles from top to bottom, then collected in a pipe attached to the bottom row of shingles. They can be adhered to any new or existing roof, and can be cut and fitted on the job to go around any type of roof projection like vent pipes, antennas, dormer windows, chimneys, etc. The units can be used to heat air, swimming pools or hot water. Manufacturer: The Solarshingles Company, Van Nuys, California.

Four different types of solar collectors are available from Sunburst: 1) solar swimming pool coils 2) aluminum and copper absorber plates, two sizes 3) boxed and glazed units and 4) all-copper solar collectors, two sizes. The transparent cover on the glazed modular collector illustrated below is Tedlar-coated fiberglass. which is impervious to breakage and to yellowing with age. Manufacturer: Sunburst Solar Energy, Menlo Park, California.

CUTAWAY OF GLAZED MODULAR COLLECTOR

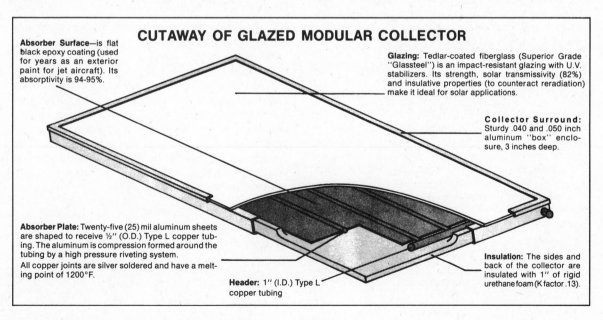

Absorber Surface—is flat black epoxy coating (used for years as an exterior paint for jet aircraft). Its absorptivity is 94-95%.

Glazing: Tedlar-coated fiberglass (Superior Grade "Glassteel") is an impact-resistant glazing with U.V. stabilizers. Its strength, solar transmissivity (82%) and insulative properties (to counteract reradiation) make it ideal for solar applications.

Collector Surround: Sturdy .040 and .050 inch aluminum "box" enclosure, 3 inches deep.

Absorber Plate: Twenty-five (25) mil aluminum sheets are shaped to receive ½" (O.D.) Type L copper tubing. The aluminum is compression formed around the tubing by a high pressure riveting system.
All copper joints are silver soldered and have a melting point of 1200°F.

Header: 1" (I.D.) Type L copper tubing

Insulation: The sides and back of the collector are insulated with 1" of rigid urethane foam (K factor .13).

Collectors and Systems

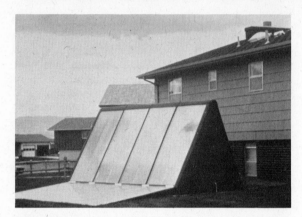

This Solar Furnace is designed for use with existing homes, new construction, small commercial buildings and mobile homes. A major advantage of the auxilary system is that it requires no major structural changes to the building for installation. Connected to the existing ductwork, the furnace works in conjunction with the existing forced air furnace. A solid-state controller contained inside the furnace and connected by electrical wiring to the house furnace decides what happens when the house thermostat calls for heat. The control determines whether the Solar Furnace, your house furnace or both will come on to satisfy the thermostat's demand for heat. Manufacturer: Champion Home Builders, Dryden, Michigan.

Solar flat plate collectors by Energy Systems are designed and tested for direct connection in domestic potable hot water systems at full utility service pressure. No heat exchanger or special water treatment is required. Freeze protection is provided without the use of anti-freeze by any one of a number of automatic control systems available. The collectors may be used for space heating, domestic water heating and swimming pool and spa heating. Manufacturer: Energy Systems, San Diego, California.

Heliotrope General offers a complete line of differential temperature thermostats for solar hot water and solar space heating control which encompasses 24 different UL listed models. They also supply an electronic thermometer for remote sensing. Manufacturer: Heliotrope General, Spring Valley, California.

Solarmaster now offers a low-cost solar collector kit that you can assemble in approximately 30 minutes. The collector measures 4' x 7'10", with all-copper waterways that snap on, and aluminum heat absorbing fins. Modular installation allows you to add more panels at any time. The firm produces a full line of solar collectors for hot water, heating, cooling, and swimming pool use. Manufacturer: Solarmaster, Santa Maria, California.

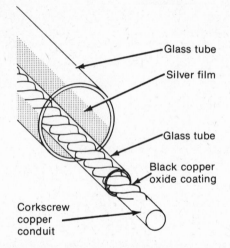

This glass tube concentrator consists of a number of parallel concentrator elements built into a frame of lightweight architectural aluminum and covered with a skin of Tedlar. Lightweight glass tubing, similar to that in fluorescent lights, forms the outer surface of the concentrator elements. Half of the outer tube surface is silvered, creating a mirror finish that concentrates the available sunlight on a spiral or corkscrew copper conduit running the length of the glass tube. This copper conduit, coated with a black copper oxide, transfers heat to water running through the conduit. Another glass cylinder fits snugly over the copper conduit, reducing convection heat losses. The collectors are compatible with off-the-shelf components such as piping, storage and controls. An important feature of the unconventional design is that, by rotating the collector's elements, the collector can be matched to the pitch of the roof before installation. Thus the collector can be mounted dead flat or perpendicular. Manufacturer: KTA Corporation, Rockville, Maryland.

Sun-A-Matic Solar Collector Systems offers free this new brochure on their flat plate collector and variable flow controller, for domestic hot water, comfort heating and swimming pools. For a copy of this brochure write: Sun-A-Matic Division, Butler Ventamatic Corp., Box 728, Mineral Wells, Texas 76067.

The Reynolds Aluminum solar collector is made with integrally-finned extruded aluminum tube, serpentined into the configuration of a flat-plate collector. The standard collector is 4' x 8' x 3⅝" and weighs 67.8 pounds when the tubes are full of liquid. The collectors are designed to operate in closed systems with inhibited water. They may be adapted to conventional systems for space heating, household or commercial hot water, swimming pool heating and combinations of these applications. Manufacturer: Reynolds Metal Company, Richmond, Virginia.

SOLAR STORAGE TANK
Model STJ

This solar hot water storage tank is one of two products A. O. Smith has announced for use in solar heating applications. The other product is a solar electric heater, similar in appearance to the storage tank except that it has only one solar connection (on the bottom) instead of two. The electric heater is designed for systems incorporating solar heating which require some backup heating system. Both products come in 30 through 120 gallon capacity sizes and have the same kind of double-density insulation found in the company's Conservationist heaters. Manufacturer: A. O. Smith Corporation, Kankakee, Illinois.

The Model 3002 Acurex Concentrating Collector is a reflecting parabolic trough solar collector designed to heat liquids or gases to temperatures between 140°F and 350°F. Typical applications include water heating, air heating, steam generation and space cooling. The collector is assembled in modules 10 feet in length; four modules are normally coupled together to form a line of collectors which is driven by a single drive system at the middle of the line. The heating fluid in the receiver tube can be water, organic liquid, or air. A high temperature concentrating collector designed for temperatures between 250°F and 600°F is also available. Manufacturer: Acurex Aerotherm, Mountain View, California.

121

Collectors and Systems

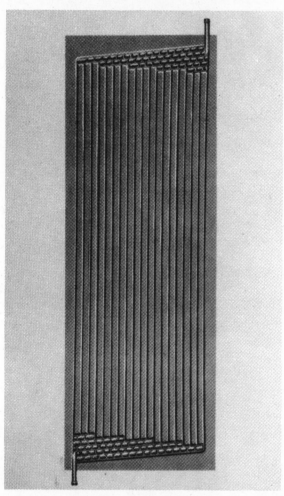

Solar-Bond panels provide integral tubes and headers within a metal plate — either aluminum or copper. They are produced by metallurgically bonding two sheets of metal together and then expanding them in selected unbonded areas to form integral flow passages. The heat transfer medium, either liquid or air, flows through the passages. The absorbers are designed for use in flat plate and concentrating collectors and as secondary heat exchangers. The company does not manufacture complete collections or total solar systems. Manufacturer: Olin Brass, East Alton, Illinois.

The "Decade 80 Solar House" in Tucson, built by the Copper Development Association, uses this combination copper roofing and solar energy panel to get up to 97% of its heating and most of its cooling from the sun's energy. The copper roofing and energy panels consist of 2' x 8' copper sheets laminated to plywood combined with rectangular copper tubes to carry the energy system's transport and storage medium, water. The copper collector-roofing system is produced by Revere Copper and Brass. For additional information, contact the Copper Development Association, New York, New York.

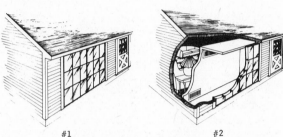
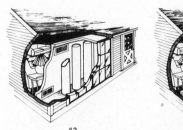
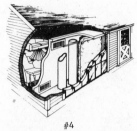

#1   #2   #3   #4

These drawings show a passive solar furnace by Kalwall called the Solar Battery System. The outside "weatherwall" (drawing #1) has a solar window that transmits 77% of the solar energy and insulates with a U equal to .40. A small "Hot Box" enclosure (#2) is highly insulated on three walls, the floor and the ceiling, with air vents installed. A row of collector-storage tubes (#3) is positioned close to, and along the entire length of the solar window wall. The more tubes used, the more heat storage capacity. These tubes are commonly filled with water, but may have other storage media. Movable insulation (#4) located between the tubes and the solar wall provides control of sun's insolation and heat loss. Natural convection currents move the air around the tubes and out of the solar furnace through the air vents installed in the partition. A fractional horsepower blower can be installed and controlled by a thermostat to increase the rate of air and heat flow. The output of this blower can be connected directly to the existing hot air system ducting. Manufacturer: Kalwall Corporation, Manchester, New Hampshire.

Prefabricated modular solar building systems can be used to solar heat patio enclosures, home additions, pool enclosures, solariums or greenhouses. The solar garden structure supplies supplemental heat and humidity to homes, while providing stored solar heat to keep the greenhouse warm at night and during cold weather. Heat is drawn into attached homes with fans, natural air circulation, or through a conventional hot air heating system. Both passive and active heat storage systems are available. Manufacturer: Solar Technology Corporation, Denver, Colorado.

Collectors and Systems

The Solar Aire Furnace is designed for immediate use, without modifying the existing home, or it may be built into new homes without radically affecting the traditional appearance. Three models are available, with rock storage capacities of 9, 12 and 15 cubic yards. Manufacturer: SunSaver Corporation, N. Liberty, Iowa.

SolarStrip is a continuously electroplated selective black chrome surface for solar collectors. Tests indicate the product offers a combination of high absorptivity (.93 min.) and low emissivity (.10 max.) among available coatings. It can be adhesively bonded to aluminum extrusions and steel fins, or spiral wrapped on tubing. Black chrome can also be formed and soldered without damage to the coating. Manufacturer: Berry Solar Products, Edison, New Jersey.

The Sunbearer evacuated tube solar collector is a high temperature collector that can provide solar heat for baseboard hot water systems. The circulating water is heated to 185°-200°F, temperatures most flat plate collectors cannot provide. The collector can also supply water hot enough to operate absorption chillers at top efficiency during the summer for air conditioning. Manufacturer: Solar Industries, Plymouth, Connecticut.

A new insulated solar energy collector cover panel system called "Sunwall" is said to be up to 50% more insulating than glass collector covers. The panel's energy transmission is 77% and its insulation U-factor is .40 — equal to 16 inches of concrete. It is suitable for use with most air or water fluid systems, or passive heating designs. The panels are produced in four to five foot widths, in any desired length up to 20 feet. Construction is shatter-proof fiberglass bonded to two sides of an aluminum grid core with a 2¾" insulating air space. Manufacturer: Kalwall Corporation, Manchester, New Hampshire.

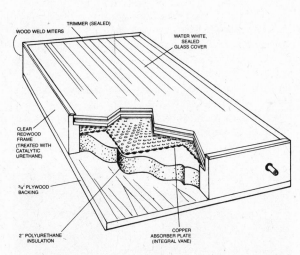

WOOD WELD MITERS
TRIMMER (SEALED)
WATER WHITE, SEALED GLASS COVER
CLEAR REDWOOD FRAME (TREATED WITH CATALYTIC URETHANE)
3/8" PLYWOOD BACKING
2" POLYURETHANE INSULATION
COPPER ABSORBER PLATE (INTEGRAL VANE)

The cut-away drawing illustrates the construction of flat plate collectors manufactured by Solar King. Each collector is constructed with redwood, stained and treated with catalytic urethane for extra protection. Two sheets of sealed water white crystal are tempered for strength and stippled for extra radiation collection (the stippled surface is designed to collect the sun's radiation from as low as 15 degrees to the horizon). The backing is 3/8" cured plywood. Manufacturer: American Solar King Corporation, Waco, Texas.

Solar greenhouse kits come complete with factory-finished modules of redwood and glass (or acrylic). All parts and tools necessary for complete assembly are shipped directly to you. The firm's literature indicates that assembly and installation can usually be completed in less than an afternoon. Manufacturer: The Energy Factory, Fresno, California.

The Sundance II solar collector features bronze anodized framing throughout and a removable safety glass cover that facilitates service while installed. An integral inverted lip on the collector frame simplifies built-in flashing applications. The collector panel measures approximately 4' x 8' and weighs 129 lbs. empty, 136 lbs. full. Recommended usage: swimming pools, spas, hot tubs, domestic water heating, space heating, and air conditioning. Manufacturer: Easton Corporation, San Jose, California.

The GED-SOLARTRON solar energy system captures radiant energy from the sun, absorbs it as energy into highly conductive materials that transfer the heat to water held in a storage tank. The heated water is then circulated through a radiator or coils while fans blow air across them into the building, thus heating the air entering the building. The system can be installed in conjunction with or independent of your existing air handler system. With the supplied directions, installation can be made by a licensed plumber, heating contractor or the home handyman. The water tank will store up to 5 million BTU's of energy that can be used at night and extended periods of limited sunshine. If necessary, a two stage thermostat will activate your present heating system or a small conventional electric, gas or oil back-up unit for extraordinarily long periods of limited sunshine or when excessive heating is required. Manufacturer: General Energy Devices, Clearwater, Florida.

Three basic types of Sun-Grabber collectors are manufactured by R-M Products: water collectors, air collectors, or combination air and water collectors. All are made in standard 24 inch nominal widths and can be any length from 4 ft. to 25 ft. long. The panels may be hoisted into position, with the glass in place, without danger of breakage. A roof deck is not necessarily required for support, since the collectors may be supported simply by roof joists or rafters at 24 inches on center. As a result, roof deck may be omitted where the collectors are located, saving in construction costs. The firm also produces two types of domestic hot water heating systems, swimming pool collectors, and a packaged air handling unit for a solar air system. Manufacturer: R-M Products, Denver, Colorado.

Rho Sigma's manufacturing and engineering departments are devoted exclusively to the development and production of solar energy controls and instruments. Their standard product line provides a wide variety of controls currently available. Their engineering efforts are directed toward system designs submitted by other manufacturing firms as well as systems conceived in-house. Manufacturer: Rho Sigma, Inc., North Hollywood, California.

Collectors and Systems

Sol-Con II is an enclosed flat plate collector utilizing the solar oven effect. Extruded aluminum fins are bonded to parallel copper tubes manifolded at each end. An extruded aluminum, self-supporting frame securely holds the backing (insulating panel), heat exchanging unit, and the front safety glass covering — with all connecting manifolds extending 2" beyond the frame on all corners. The collector may be installed in the widest variety of configurations. Manufacturer: Solar Concepts, San Jose, California.

The Solar, Inc. heat storage system is a patented storage concept that uses eutectic salts. This material stores almost five times the heat of an equal volume of water and twenty five times the heat of an equal volume of rock. The firm's heat storage trays are designed to be self stacking and are permanently sealed with the eutectic inside. The trays should never need replacement or recharging. Manufacturer: Solar, Inc., Mead, Nebraska.

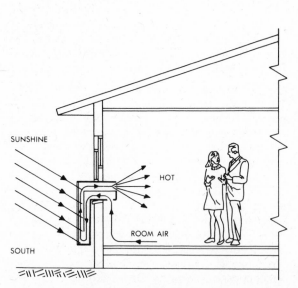

The ISIS Solar Heater is a unique active solar energy system mounted on a vertical wall using air as the working fluid. The ISIS is an auxiliary heater and is not designed to supply all of the heating needs of an average house. It will, however, supply sufficient heat for a 16' x 16' exterior room with a south facing wall during days when the sun is shining. Radiant energy from the sun passes through the collector cover and is absorbed by the special heat transfer elements which fill the panel. When the internal panel temperature reaches a pre-set limit, a thermostat automatically turns on a small blower which draws cool inside air into the panel and around, through, and over the heat transfer elements warming the air and blowing it into the room. The panel continues to operate as long as the sun shines on it. The blower, differential thermostat and controls are housed in a transition piece from the panel and protrude into the room no more than a small window air conditioner. The 3' x 7' panel weighs less than 75 pounds, which makes it easy to install for the heating season or easy to store when no heating is required. Manufacturers: Decker Manufacturing, Keokuk, Iowa.

This new home employs a solar-assisted heat pump system which combines the energy-efficient heat pump with a solar heating system. Components of the system include a Fedders Flexhermetic II split system heat pump, a duct-mounted solar hydronic heating coil, a control panel, a thermal storage tank and a bank of outdoor solar collectors. At the heart of each solar collector is a black absorber plate covered with one or two layers of glass and/or plastic. The absorber plate collects energy from the sun and uses it to heat water in the storage tank. From this point on, the system functions as a conventional fan-coil hot water heating system. Manufacturer: Fedders Corporation, Edison, New Jersey.

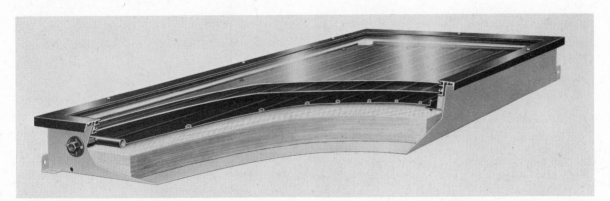

The LSC18-1 flat-plate collector is the development of solar design and manufacturing by Lennox and the solar research group by Honeywell, Inc. Features include a high performance absorber plate with exclusive black-chrome coating application, complete isolation of absorber plate to reduce conductivity heat loss, and double glass, acid etched for anti-reflection to increase solar transmission. Manufacturer: Lennox Industries, Marshalltown, Iowa.

Collectors and Systems

The illustrated catalog of Solar Energy Products describes a varied product line, including solar collectors, cover plate options, absorber plates, heat exchanger, pumps, differential temperature controllers, fluid handling packages, storage subsystems, mounting systems, speciality hardware, and pre-packaged solar water heating kits. The catalog also describes the combinations of optional equipment available to the design engineer to coordinate system performance according to regional requirements. Manufacturer: Solar Energy Products, Gainesville, Florida.

The Sunstream 60 Solar Collectors on this Vermont home feature baked-white enamel aluminum frames and acrylic covers. Absorber plates are offered in all-aluminum rollbond or proprietary Finplank design, which features copper fluid passages mechanically locked between extruded aluminum planks. The arch of the collectors provides a distinctive appearance. Manufacturer: Grumman Sunstream, Ronkonkoma, New York.

Hot Water Systems

The Solcan solar hot water heating system is designed to supply the needs of a family of four (approximately 120 gallons per day per household, assuming sunshine is average). The system is available in self-installation kit form, complete with solar collectors, frames, pumps, motors, tank and electronic controls. Manufacturer: Albuquerque Western Solar Industries, Albuquerque, New Mexico.

The basic function of a typical flat plate solar collector for heating domestic hot water is explained in this drawing. As solar energy is absorbed into the collector, an anti-freeze solution is heated and moves to a heat retention tank. When needed, this heated fluid is pumped from the tank into the home's conventional hot water heating system and eventually is circulated back through the collector to be heated again. Drawing courtesy of Libbey-Owens-Ford Company.

SERC'S domestic hot water solar pre-heat kit connects in series with your present hot water tank and will heat 40 gallons up to 140°F. Two 14 sq. ft. collectors mount on a roof or exterior wall and collect and convert solar radiation. Kits come complete with lightweight copper collectors, pre-heat tank with factory installed copper heat exchanger, stainless steel pump and solid state control system. Manufacturer: Solar Energy Research Corporation, Longmont, Colorado.

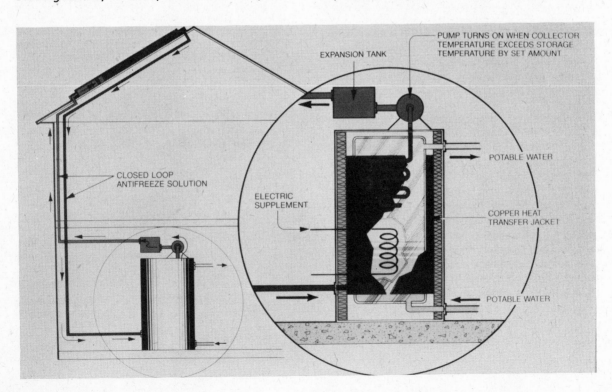

CLOSED LOOP ANTIFREEZE SOLUTION

EXPANSION TANK

PUMP TURNS ON WHEN COLLECTOR TEMPERATURE EXCEEDS STORAGE TEMPERATURE BY SET AMOUNT

ELECTRIC SUPPLEMENT

POTABLE WATER

COPPER HEAT TRANSFER JACKET

POTABLE WATER

Hot Water Systems

A complete solar domestic hot water system that can supply up to 100 percent of the hot water used in a home is being marketed by Lennox Industries. The system, known as the LSHW1 series, is composed of flat-plate solar collectors and a solar hot water module. These combined with a new or existing conventional hot water heater comprise all the major components of the system. The system is suitable for new construction or retrofit. The hot water module (shown up above) acts as a solar heat storage tank, a heat exchanger (double-walled) and a control center all in one. The system's controls, pump, valves and plumbing connections are factory mounted on top of the module for ease of installation and convenience in servicing. The series consists of five basic systems, differing only in the number of collectors used and the size of the solar storage module (available in 66, 82 and 120 gallon capacity). These five choices allow the homeowner to choose the percentage of hot water the solar system will provide. Manufacturer: Lennox Industries, Marshalltown, Iowa.

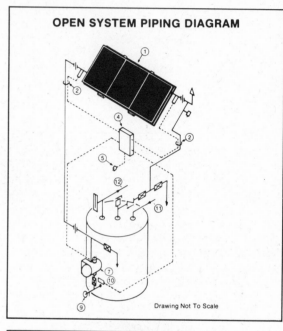

OPEN SYSTEM PIPING DIAGRAM

Drawing Not To Scale

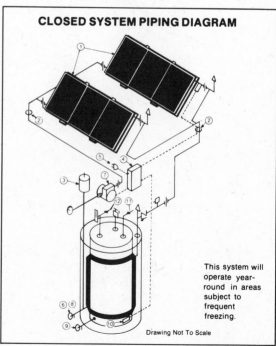

CLOSED SYSTEM PIPING DIAGRAM

This system will operate year-round in areas subject to frequent freezing.

Drawing Not To Scale

Solar Energy Products (at left) offers two types of solar water heating systems: open and closed loop. In the open loop system, water is pumped from the storage tank to the panels where it absorbs heat, then returns to the tank for storage. This function is controlled by a differential temperature controller which allows the water to circulate between the panels and the tank only when the panels are at a higher temperature than the tank. This results in pump activation only when the system can realize a net gain. In the case of inclement weather conditions, the back-up electric element may be activated to provide uninterrupted supply of hot water. The system is protected from freezing by either a recirculating or drain feature. When freezing conditions exist, panels and exposed piping are either warmed with recirculated water or drained. While the open loop system is the most economical of solar systems, the closed loop system does have important advantages — the system is freeze protected in the coldest climates, the panel fluid loop is free from corrosion, and there is no possibility of mineral build-up in the panels if harsh water conditions exist. The closed loop system includes panels, tank with a built-in heat exchanger and a fluid handling package. A heat transfer fluid is pumped between the panels and the heat exchanger tank. The exchanger transfers the heat from the fluid to the potable water inside the storage tank. The heat exchanger tank is a double-wall type, which is safe from heat transfer fluid contamination. Manufacturer: Solar Energy Products, Gainesville, Florida.

This solar hot water system by Raypak is located on a slope behind a single family residence and has been designed to blend in with the architecture and landscaping. Raypak offers a complete line of solar water heating products, from single and double glazed collector panels to complete package systems for residential or commercial use. Residential systems are available for new construction as well as for retrofit on existing buildings. Other systems for non-freezing climates are available, as are systems incorporating a simplified thermosyphon design. Manufacturer: Raypak, Westlake Village, California.

The chart and calculations are from a brochure by Revere Copper and Brass on their Sun-Pride solar energy domestic hot water system.

HOT WATER USAGE AND SUGGESTED NUMBER OF COLLECTORS

PERSONS	AVERAGE HOT WATER REQUIREMENTS (GALLONS)	STORAGE TANK (GALLONS)	NUMBER OF COLLECTORS		
			EXCELLENT SOLAR AREA	GOOD SOLAR AREA	FAIR SOLAR AREA
2	40	66	2	2	2
3	60	66	2	2	3
4	80	82	2	2 or 3	3
5	100	100	3	3	4
6	120	120	3	3	4

This is based on an average family usage of 20 gallons of hot water per day per person and the average production of 1⅓ gallons of hot water per square foot of collector per day.

Hot Water Systems

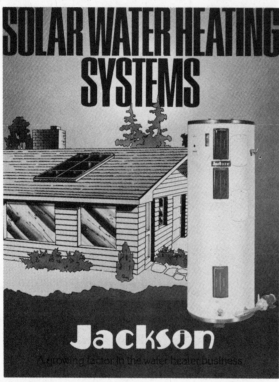

Literature available from W. L. Jackson Manufacturing explains and illustrates the firm's energy-saving solar water heating system. The system consists of two basic subassemblies: (a) the rooftop solar collector panel(s) and (b) the solar water heater storage tank. Both of these basic subassemblies are connected by piping to form a "solar loop." The loop carries the heat transfer fluid, which is circulated by a pump, up to the solar collector panel located on the roof, down through the solar water heater tank, and back to the rooftop collector. The system's circulating pump will only activate when there is enough solar energy to heat the water. On extended sunless days, an auxiliary electric heating element located in the upper one third of the storage tank is automatically activated to provide approximately 30 gallons of hot water per hour. Manufacturer: W. L. Jackson Manufacturing, Chattanooga, Tennessee.

The Ford solar water heater consists of an insulated, stone-lined, hot water storage tank with a pancake coil of finned copper tubing in the bottom for heating the water. The tank is equipped with a surface mounted thermostat to control the operation of a circulator. When the thermostat calls for heat, the fluid from the solar panels is circulated through the coil to generate hot water. A differential thermostat can be field substituted. It is also available with an electric element for heating the water during extended periods of cloudy weather. Because the tank is completely stone-lined, it is possible to use a copper heat exchanger with a large surface area in the tank without creating a condition which would cause tank failure from electrolytic corrosion. Manufacturer: Ford Products Corporation, Valley Cottage, New York.

Solar Development's hot water heater is uniquely designed to be compatible with existing household and commercial plumbing systems. Instead of utilizing a separate hot water tank that circulates water by the thermal siphon principle, the company uses a standard hot water tank and forces circulation to speed up heat transfer and system effectiveness. Shown above is the dual mounting system for two 2' x 10' hot water collectors. The system's advantages include only four roof mounting points and a low profile. Manufacturer: Solar Development, West Palm Beach, Florida.

Model TC 65E

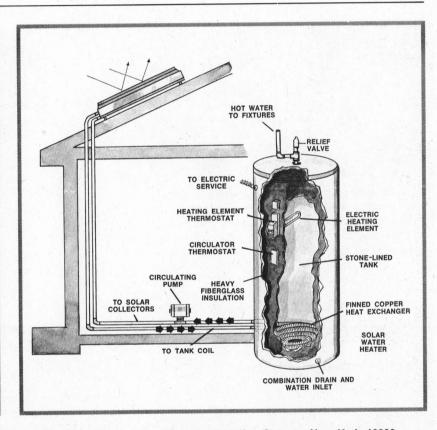

HOT WATER
TO FIXTURES

RELIEF
VALVE

TO ELECTRIC
SERVICE

HEATING ELEMENT
THERMOSTAT

ELECTRIC
HEATING
ELEMENT

CIRCULATOR
THERMOSTAT

STONE-LINED
TANK

CIRCULATING
PUMP

HEAVY
FIBERGLASS
INSULATION

TO SOLAR
COLLECTORS

FINNED COPPER
HEAT EXCHANGER

SOLAR
WATER
HEATER

TO TANK COIL

COMBINATION DRAIN AND
WATER INLET

FORD PRODUCTS CORPORATION, Ford Products Road, Valley Cottage, New York 10989

The GED-Solartron collector for solar water heating is a compact, modular unit, completely sealed in a durable fiberglass case and mounted on the roof. The collector is connected through concealed piping to a counter-flow heat exchanger tank or an ancillary counter-flow insulated heat exchanger unit in conjunction with your existing hot water tank. Total power consumption is 100 watts intermittently. All fittings and components are standard sizes and materials to permit easy connection to your existing plumbing and to conform to standard building codes. Easy to follow installation directions are included with each unit for the licensed plumber or the home handyman. Manufacturer: General Energy Devices, Clearwater, Florida.

Hot Water Systems

Sunearth solar collectors for hot water heating come in two basic models. The first is a modular skylight mounted type (field assembly required). The second is a fully assembled box type flat plate collector for mounting on existing roofs or frames. Both collectors use water or anti-freeze solutions as the heat transfer medium. Special features include double glazing consisting of heat, scratch and craze-resistant acrylic and heat-proof Teflon film. Domestic hot water systems also come in two types of completely automatic packages. Both systems come complete with collectors, tank, controls, fittings and instructions. Manufacturer: Sunearth Solar Products Corporation, Green Lane, Pennsylvania.

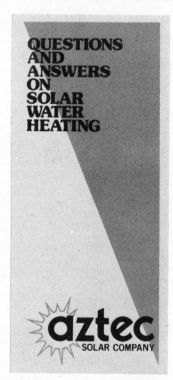

QUESTIONS AND ANSWERS ON SOLAR WATER HEATING

aztec
SOLAR COMPANY

This free folder poses and answers 14 common questions about solar hot water heating, from how large a hot water tank is needed to how much maintenance is required. The firm also provides system design, greenhouse heating, collectors, controllers and components. Manufacturer: Aztec Solar Company, Maitland, Florida.

This solar hot water heating system is ground installed for applications where roof mounting is impractical. It's designed for a family of 3 or 4 with usual appliances and up to 1½ baths. The system includes a non-corrosive solar collector and cover, mounted on a lightweight steel frame; an 84 gallon insulated Therma Tank with built-in copper coil heat exchanger; differential controller; circulating pump; piping between collector and tank; and aluminum-clad siding enclosure (not shown). Split systems for roof mounting are also available. Manufacturer: Energy Absorption Systems, West Sacramento, California.

Installation of a solar domestic hot water system by Sun-A-Matic Solar Div., Butler Ventamatic Corp., Mineral Wells, Texas.

Locating southern best position for collector, using compass

Laying out to locate roof mount

Marking roof before drilling holes to set up roof mount

Setting collector in position

Installing piping to collector

Setting domestic hot water tank

Swimming Pool Heating

The Solar-Pak solar pool heater can be installed on almost all existing pools, and on all new pools, with no additional equipment required other than distribution piping and mounting hardware. The metal panels can be located on a roof, patio cover, hillside, or frame adjacent to the pool. Two convenient control system options are available — a manual system for the homeowner who wishes to monitor and adjust the solar heater himself, or a fully automatic control system that provides complete operating convenience and optimum energy utilization. Manufacturer: Raypak, Westlake Village, California.

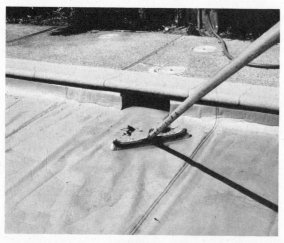

The Heat Saver insulated pool blanket floats on the surface of the water to insulate against heat-robbing night air. A suction-like bond is created between the blanket and the water which holds it down even during strong winds — no tie-downs are required. Leaves and debris which collect on the cover can simply be brushed or hosed off into the pool filter inlets for removal. The blanket folds into a small bundle for storage and is easy to handle — a 500 square foot Heat Saver weighs about 30 pounds. Standard sizes range from 12' x 24' to 26' x 45'. Custom sizes for irregularly shaped pools can also be ordered. Manufacturer: MacBall Industries, Oakland, California.

Alcoa's solar heating panels for outdoor swimming pools mount on a wood or aluminum frame and are installed in a sloping position, either on the ground or on the roof of a house, pool building, garage or covered patio. The system is simple, consisting of a series of black solar heating-collecting panels, piping and controls — all tied into the existing pool filtering system. Should the temperature of the solar heater panels rise four degrees above the temperature of the pool water, a sensor reacts and a solenoid valve diverts pool water into the panels. As water circulates through the panels, it absorbs solar energy, is warmed steadily and then returned to the pool. After water temperatures reach a comfortable level, the solar components shut off until needed again. A manual control system is standard for the homeowner who desires to monitor heating and make adjustments himself. Manufacturer: Aluminum Company of America, Pittsburgh, Pennsylvania.

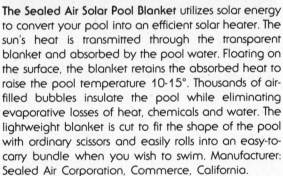

The Sealed Air Solar Pool Blanket utilizes solar energy to convert your pool into an efficient solar heater. The sun's heat is transmitted through the transparent blanket and absorbed by the pool water. Floating on the surface, the blanket retains the absorbed heat to raise the pool temperature 10-15°. Thousands of air-filled bubbles insulate the pool while eliminating evaporative losses of heat, chemicals and water. The lightweight blanket is cut to fit the shape of the pool with ordinary scissors and easily rolls into an easy-to-carry bundle when you wish to swim. Manufacturer: Sealed Air Corporation, Commerce, California.

A trained 2 man crew can install a complete residential pool system by Sunburst in 2 to 3 hours. The competent handyman may take a bit longer, but quick-connect couplings and coil support racks require no special skills or uncommon tools. The black plastic used in the coils is chemically treated to inhibit ultraviolet degradation and is designed to carry water at domestic water pressures of 65 psi and more. A total coil area of roughly 50-60% of the swimming pool surface area is adequate for most pools. Manufacturer: Sunburst Solar Energy, Menlo Park, California.

A number of adventurous, budget-conscious home-owners have installed solar heating equipment without the benefit or expense of professional help. Exemplifying this do-it-yourself approach is Paul H. Nankivell, Los Angeles, who recently built a solar heater for his 40,000-gallon pool in the Hollywood Hills. Mr. Nankivell, whose background is in marketing not engineering, built his pool solar heater entirely by himself in about a week's leisure time. "I built my own pool solar heater chiefly to acquaint myself with this important new market now opening up in California and the rest of the country," explains Nankivell, who is a major manufacturer's representative for the plumbing, heating, and cooling industries. "Solar collectors, after all, are a relatively simple operation." Nankivell's solar pool heater consists of four 4x8-foot collectors made of Type M copper tube on 16-ounce copper sheet mounted on particleboard. His chief aide in this do-it-yourself project, he says, was the 46-page booklet, "How to Design and Build a Solar Swimming Pool Heater," available free by writing to the Copper Development Association, 405 Lexington Avenue, New York, New York 10017.

Manufacturers' Index

A 1 Prototype
1288 Fayette St.
El Cajon, CA 92020

Acurex Aerotherm
485 Clyde Ave.
Mountain View, CA 94042

Albuquerque Western Solar Industries
612 Comanche, NE
Albuquerque, NM 87107

Aluminum Company of America
Alcoa Building
Pittsburgh, PA 15219

American Solar Heat Corp.
7 National Place
Danbury, CT 06810

American Solar King Corp.
6801 New McGregor Highway
Waco, TX 76710

American Solar Systems
415 E. Branch St.
Arroyo Grande, CA 93420

Aztec Solar Company
Box 272
Maitland, FL 32751

Berry Solar Products
Woodbridge at Main
Box 327
Edison, NJ 08817

Champion Home Builders Co.
5573 E. North
Dryden, MI 48428

Copper Development Association
405 Lexington Ave.
New York, NY 10017

Decker Manufacturing
312 Blondeau
Keokuk, IA 52632

del Sol Control Corp.
11914 U.S. 1
Juno, FL 33408

E & K Service Company
16824 - 74th NE
Bothell, WA 98011

Easton Corporation
23773 McKean Rd.
San Jose, CA 95141

Edmund Scientific Co.
101 E. Gloucester Pike
Barrington, NJ 08007

Energy Absorption Systems, Inc.
860 S. River Rd.
West Sacramento, CA 95691

The Energy Factory
5622 E. Westover, Suite 105
Fresno, CA 93727

Energy Systems
4570 Alvarado Cyn Rd.
San Diego, CA 92120

Entropy Limited
5735 Arapahoe Ave.
Boulder, CO 80303

Fafco, Incorporated
138 Jefferson Dr.
Menlo Park, CA 94025

Fedders Corporation
Edison, NJ 08817

Ford Products Corporation
Ford Products Road
Valley Cottage, NY 10989

General Energy Devices
Clearwater, FL 33515

Grumman Sunstream
4175 Veterans Memorial Hwy.
Ronkonkoma, NY 11779

Heliotrope General
3733 Kenora Dr.
Spring Valley, CA 92077

Hitachi Chemical Co.
437 Madison Ave.
New York, NY 10022

Ilse Engineering
7177 Arrowhead Rd.
Duluth, MN 55811

W.L. Jackson Mfg. Co.
Box 11168
Chattanooga, TN 37401

KTA Corporation
12300 Washington Ave.
Rockville, MD 20852

Kalwall Corporation
1111 Candia Road
Manchester, NH 03103

Lennox Industries
350 S. 12th Ave.
Marshalltown, IA 50128

Libbey-Owens-Ford Co.
811 Madison Ave.
Toledo, OH 43695

Lof Bros. Solar Appliances
1615 17th St.
Denver, CO 80202

MacBall Industries
5765 Lowell St.
Oakland, CA 94608

Olin Brass
East Alton, IL 62024

Owen Enterprises
436 No. Fries Ave.
Wilmington, CA 90744

PPG Industries, Inc.
Solar Systems Sales
One Gateway Center
Pittsburgh, PA 15222

R-M Products
5010 Cook St.
Denver, CO 80216

Raypak, Inc.
31111 Agoura Rd.
Westlake Village, CA 91359

Revere Copper and Brass
605 Third Ave.
New York, NY 10016

Reynolds Metals Co.
Richmond, VA 23261

Rho Sigma, Inc.
11922 Valerio St.
North Hollywood, CA 91605

Robertshaw Controls Co.
100 W. Victoria St.
Long Beach, CA 90805

Sealed Air Corporation
2015 Saybrook Ave.
Commerce, CA 90040

A.O. Smith Corporation
Kankakee, IL 60901

Solar, Incorporated
Box 246
Mead, NB 68041

Solar Corporation of America
100 Main St.
Warrenton, VA 22168

Solar Concepts
818 Charcot Ave.
San Jose, CA 95131

Solar Development, Inc.
4180 Westroads Dr.
West Palm Beach, FL 33407

Solar Energy Products, Inc.
1208 NW 8th Ave.
Gainesville, FL 32601

Solar Energy Research Corp.
701B South Main St.
Longmont, CO 80501

Solar Energy Systems, Inc.
One Olney Ave.
Cherry Hill Industrial Center
Cherry Hill, NJ 08003

Solar Industries, Inc.
Box 303
Rt. 6 and Burr Rd.
Plymouth, CT 06782

Solar Innovations
412 Longfellow Blvd.
Lakeland, FL 33801

Solar Manufacturing Co.
40 Conneaut Lake Rd.
Greenville, PA 16125

Solar Power Supply
Rt. 3, Box A10
Evergreen, CO 80439

Solar Systems Division
12400 49th St. South
Clearwater, FL 33520

Solar Technology Corp.
2160 Clay St.
Denver, CO 80211

Solar West, Inc.
Box 892
Fresno, CA 93714

Solarator
Box 277
Madison Heights, MI 48071

Solarmaster
722-D West Betteravia Rd.
Santa Maria, CA 93454

Solaron Corporation
300 Galleria Tower
720 South Colorado Blvd.
Denver, CO 80222

The Solarshingles Co.
14532 Vanowen St.
Van Nuys, CA 91405

Solarsystems Industries, Ltd.
5511 - 128th St.
Surrey, B.C., Canada V3W 4B5

Sun-A-Matic Division
Butler Ventamatic Corp.
Box 728
Mineral Wells, TX 76067

Sunburst Solar Energy
123 Independence Dr.
Menlo Park, CA 94025

Sunearth Solar Products Corp.
R.D. 1, Box 337
Green Lane, PA 18054

SunSaver Corporation
Box 276
N. Liberty, IA 52317

Zomeworks Corporation
P.O. Box 712
Albuquerque, NM 87103

Sources

Additional Reading

Alternative Natural Energy Sources in Building Design. Albert J. Davis and Robert P. Schubert, Van Nostrand Reinhold, New York, 1974. 111 pp.

The Coming Age of Solar Energy. D. S. Halacy, Avon Books, New York, 1975. 248 pp.

The Complete Solar House. Bruce Cassiday, Dodd, Mead & Co., New York, 1977. 212 pp.

Designing & Building a Solar House. Donald Watson, Garden Way, Charlotte, Vermont, 1977. 282 pp.

Direct Use of the Sun's Energy. Farrington Daniels, Ballantine Books, New York, 1974. 271 pp.

Energy; the Case for Conservation. Worldwatch Institute, 1776 Massachusetts Ave. NW, Washington, D.C., 1976.

Harnessing the Sun. John H. Keyes, Morgan Press, Milwaukee, Wisconsin, 1975. 196 pp.

How to Build a Solar Heater. Ted Lucas, Ward Ritchie Press, Pasadena, California, 1975. 236 pp.

In the Bank . . . Or Up the Chimney? A Dollar and Cents Guide to Energy-Saving Home Improvements. Superintendent of Documents, U.S. Government Printing Office, Washington, D.C., Stock No. 023-000-00297-3. 72 pp.

Low-Cost, Energy-Efficient Shelter for the Owner and Builder. Edited by Eugene Eccli, Rodale Press, Emmaus, Pennsylvania, 1976. 408 pp.

New Low-Cost Sources of Energy for the Home. Peter Clegg, Garden Way, Charlotte, Vermont, 1977. 252 pp.

Solar Energy for Man. Brian J. Brinkworth, Wiley, New York, 1972. 251 pp.

The Solar Energy Handbook. Henry Landa, et al., Milwaukee, Wisconsin, 1976.

Solar Energy: Technology and Applications. J. Richard Williams, Ann Arbor Science Publishers, P.O. Box 1425 Ann Arbor, Michigan, rev. ed. 1977. 176 pp.

Solar Energy Thermal Processes. John A. Duffie and William A. Beckman, Wiley, New York, 1974. 386 pp.

Solar Home Book. Bruce Anderson and Michael Riordan, Cheshire Books, Harrisville, New Hampshire, 1976. 297 pp.

Solar Homes and Sun Heating. George Daniels, Harper & Row, New York, 1976. 178 pp.

Sun Earth. Richard L. Crowther, et al., Crowther/Solar Group Denver, Colorado, 1976. 232 pp.

Sunspots. Steve Baer, Zomeworks, P.O. Box 712, Albuquerque, New Mexico, 1975. 115 pp.

30 Energy-Efficient Houses . . . You Can Build. Alex Wade and Neil Ewenstein, Rodale Press, Emmaus, Pennsylvania, 1977. 316 pp.

A Time to Choose — America's Energy Future. Energy Policy Project of the Ford Foundation, Dorothy K. Newman and Dawn Day, Ballinger Publishing Co., J. B. Lippincott Co., 1974. 511 pp.

Your Home's Solar Potential. I. Spetgang and M. Wells, Edmund Scientific Co., Barrington, New Jersey, 1976. 6

Solar Design

Alten Associates, Inc.
3080 Alcott St., Suite 200-D
Santa Clara, CA 95051

Ames Design Collaborative
208 Fifth St.
Ames, IA 50010

Steve Baer
Zomeworks Corp.
P.O. Box 712
Albuquerque, NM 87103

Blue/Sun Ltd.
P.O. Box 118
Farmington, CT 06032

Burt, Hill & Associates
610 Mellon Bank Building
Butler, PA 16001

Lee Porter Butler
3375 Clay St.
San Francisco, CA 94118

Crowther/Solar Group
310 Steele St.
Denver, CO 80206

The Hawkweed Group Ltd.
4643 N. Clark St.
Chicago, IL 60640

Mike Jantzen
Box 171
Carlyle, IL 62231

Jespa Enterprises
P.O. Box 11
Old Bridge, NJ 08857

Doug Kelbaugh
Environment
70 Pine St.
Princeton, NJ 08540

James Lambeth, AIA
1591 Clark
Fayetteville, AR 72701

Moore, Grover, Harper, P.C.
Main St.
Essex, CT 06426

Bob Schmitt Homes
766 Gate Post Rd.
Strongsville, OH 44136

Sun House Design
P.O. Box 130
Occidental, CA 95465

Sundesigns
P.O. Box 3102
Aspen, CO 81611

Sunworks, Inc.
Guilford, CT 06437

Thomason Solar Homes, Inc.
6802 Walker Mill Rd., S.E.
Washington, D.C. 20027

Total Environmental Action, Inc.
1 Church Hill
Harrisville, NH 03450

Donald Watson
Box 401
Guilford, CT 06437

Malcolm B. Wells
Dale Ave. & N. Park Dr.
Cherry Hill, NJ 08002

Exciting home planning ideas.

Ideas for every facet of home planning, home plans in every architectural style that you can buy and build, as well as a complete guide to your home product needs and a practical solution to the mysteries of solar energy systems are yours direct from the Bantam/Hudson Planning Center.
See order form below.

am/Hudson Plans Books

tstanding collection of home
to buy, in every architectural
Designs for every section of the
y at a price you can afford.
Home Improvement Projects.
$2.95.

0 Custom Homes Plans 112 pages
onial Home Plans 112 pages
ntemporary Home Plans 112 pages
all Home Plans 112 pages
sure Home Plans 112 pages
me Improvement Project Ideas
                              80 pages

plete Home Catalog

rst comprehensive source book
ere-to-get-it information. A
cal guide to building products
meowner or professional.
les the Automatic Secretary to
u additional information on any
t covered. 350 plus pages,
+ $1.45 postage and handling.

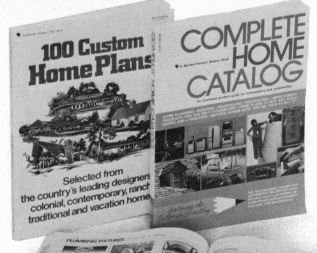

Bantam/Hudson Idea Books

A picture-packed series of elegant home planning ideas for your new home, your vacation home or that long-planned remodeling project designed to make your home a better place to live. Each $4.95.

Kitchen Ideas 112 pages
Bathroom Ideas 112 pages
Decks and Patios 112 pages
Fireplace Ideas 112 pages
Remodeling Ideas 128 pages
Bonus Rooms 128 pages
Bedroom Ideas 128 pages
Vacation Homes 128 pages

A Practical Guide to Solar Homes

All the basic information you need if you are thinking solar. Includes thirty solar and energy-conserving home plans to buy and a comprehensive listing of products on the market. 144 pages, $6.95.

How To Order Your Home Plans

- Enter plan number and number of sets wanted on order form as indicated.
- Mirror Reverse plans have all lettering and dimensions reading backwards—you will need at least two sets; one regular and one in mirror reverse.
- Materials lists and reverse plans are available only when noted in plans copy.
- Specify on order form if you want reverse plans or materials lists.

- Quality lists do not list quantities needed.
- If you plan to build, we suggest a minimum of 4–6 sets for your lender, builder, subcontractors, local building departments, etc.
- PLANS HOTLINE — You may speed your plans order by dialing TOLL FREE 800-227-8393. (California residents phone direct 415-941-6700.) For Master Charge or BankAmericard cardholders only. (Sorry, no C.O.D.'s)

Plan Prices

1 set $75; 4 sets $105; 8 sets $148; each additional set $12.

Plan	Page	Plan	Page	Plan	Page
2148	103	6603	101	6908	102
2240	113	6701	93	6909	91
2260	111	6801	85	6910	104
2341	88	6901	84	7001	94
3334	108	6902	87	7002	105
3895	96	6903	97	7003	99
6007	89	6904	95	7101	112
6501	86	6905	92	7102	110
6601	90	6906	106	7103	109
6602	98	6907	100	7301	107

Mail to: **HOME PLANS**
Hudson Home Publications
289 S. San Antonio Rd.
Los Altos, Calif. 94022

Order Form

Please send me _____ sets of
blueprints for Plan No. _____ Cost: $ _____
Postage and Handling .... $ 2.50
Allow 10 working days for delivery
California Residents add 6% Sales Tax .... $ _____
TOTAL $ _____

I hereby authorize Hudson Publishing Company to execute a sales slip on my behalf against ☐ Master Charge ☐ Visa

in the amount of $ _____ Card Exp. Date _____

Your Signature _____

Name (print) _____
Address _____
City _____
State _____ Zip _____

Make check or money order payable to Hudson Home Publications/SLR

Mail to: **HOME PLANS**
Hudson Home Publications
289 S. San Antonio Rd.
Los Altos, Calif. 94022

Order Form

Please send me _____ sets of
blueprints for Plan No. _____ Cost: $ _____
Postage and Handling .... $ 2.50
Allow 10 working days for delivery
California Residents add 6% Sales Tax .... $ _____
TOTAL $ _____

I hereby authorize Hudson Publishing Company to execute a sales slip on my behalf against my ☐ Master Charge ☐ Visa

in the amount of $ _____ Card Exp. Date _____

Your Signature _____

Name (print) _____
Address _____
City _____
State _____ Zip _____

Make check or money order payable to Hudson Home Publications (Sorry, no C.O.D.'s)